THE HUMAN SYSTEM

From Entropy to Ethics

Eulogy Edition

DICK E. HAMMOND
Doctor of Education
Retired Professor Emeritus
Texas State University

Printed in the United States of America

10 9 8 7 6 5 4 3 2 1

ISBN 0–536–57895–8
BA 2928

This book was previously published by Prentice-Hall, Inc.

MINUTEMAN PRESS®
313 E. Hutchison Street, San Marcos, TX 78666

DEDICATION

My wife Vy and I dedicate these past twenty years
of the conceptual work contained in this book,
THE HUMAN SYSTEM FROM ENTROPY TO ETHICS,
to the children we share:

Charles

Keith

Kevin

Kirby

Sha

Sue

and their five spouses, who have contributed eight
grandchildren to our gene pools. We hope we have done
our best for all of you. And we'll keep trying.

That there should be any connection between the communication of messages as studies in information theory and the theory of thermodynamics may occasion considerable surprise. The latter discipline is ordinarily thought of solely in connection with the work of heat engineers and physical chemists and has the reputation of being an extremely difficult subject, particularly among young students of physics, chemistry and engineering who rarely understand it without having their noses rubbed in it several times. Actually, it is one of the most general and all-embracing scientific theories ever invented, and in its scope can cover all the transformations of matter both living and nonliving. It is of such great significance to every human being that some acquaintance with it is imperative for every thoughtful person. Fortunately, such an acquaintance can be arrived at without the use of elaborate mathematical analysis.

Robert Bruce Lindsay, *The Role of Science in Civilization* (Westport: Greenwood Press, 1973), p. 153.

The direct relationship between entropy and randomness in any event in nature has been simply formulated in Appendix A.

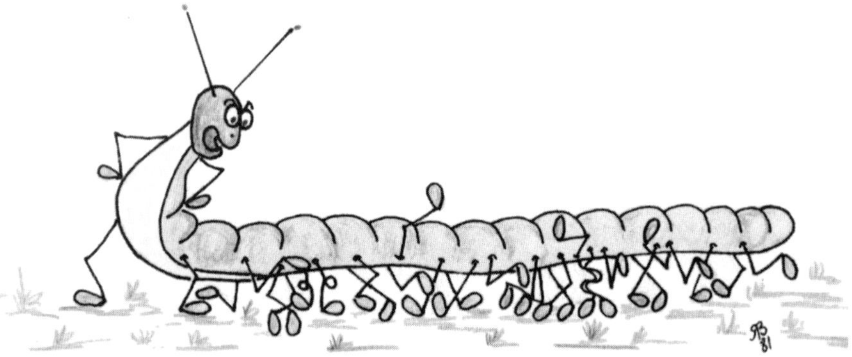

Allright you guys we ain't goin' anywhere until you reduce that ENTROPY !

ACKNOWLEDGMENTS

I wish to thank the Nobel laureate, Dr. Ilya Prigogine, for the sharing of his ideas and his writings, as well as for his encouragement in this endeavor. Also at the University of Texas at Austin, in the Ilya Prigogine Center for Research in Statistical Mechanics and Thermodynamics, Research Scientist Dr. Dilip Kondepudi was very helpful with his readings and comments on the sections dealing with systems.

I also want to thank Dr. Marguerite Gillis of the Texas State University Reading Program for her editorial suggestions on the sections dealing with information theory. Dr. William Henry in the Texas State University Secondary Education Program deserves a special thanks for reading over and making editorial comments on Parts II and III.

LaVerne DeWitt and Debbie Workman of the Texas State University Department of Curriculum and Instruction did all of the careful typing. They, too, deserve a special thanks.

For the Eulogy Edition: Carol Dolezal for the illustrations immediately preceding the new epilogue; Sabrina "Bri" Tucker (graphicdesign.com) for revisions, and Minuteman Press of San Marcos, Texas for production.

Anything in this book remaining unclear or inaccurate, is of my unaware doing.

d. e. h.

P R O L O G U E

The Human System will help educators:

1. Classify energy - matter, and why humans and their behaviors are a part of nature.

2. List the basic binding forces in the universe, and other important collectors of energy - matter.

3. Compare the laws of energy to the laws of information.

4. Know how information differs from knowledge.

5. Draw arrowed flow lines illustrating the systems analyzed, (e.g., energy / matter —> information input-output).

6. Conceptualize ***Energy Flow, Energy Flow!***
Where Does It Come From?
Were Does It Go?

7. Teach how systems are classified, and what it means to say that some are unnatural and some are natural.

8. Explain the differences between personal beliefs and scientific theories.

9. Bring order out of disorder in their systems, and their learners' behaviors.

10. Comprehend how they use the entropy ethic, and why they should.

September 2005

EULOGY EDITION

Ilya Prigogine died on May 28, 2003. Thus, I want this fourth edition to be known as the Eulogy Edition. This book is his book as much as mine. The title belongs to him as much as to me. He wanted the entropy ethic to be taught to children all over the planet.

Ilya asked his post-doctoral students and staff to help me in any way they could, and they did help many times. Ilya wanted nonequilibrium thermodynamics in all educational curricula.

I was finishing my doctorate the year Ilya won the Nobel Prize in 1977. My dissertation, *An Analysis of Entropy Reduction and Its Implications for Ethical Instruction in Public Education,* was approaching what he wanted to see happen in education. I promised him I would devote my life to teaching the laws of thermodynamics with the laws of information.

This brilliant, happy man taught us how to put the pieces of existing knowledge together and make them one...NATURE. I have known his happiness. And, I know my thousands of students are going through a more joyful life, as I have, because of Ilya Prigogine.

June 1, 2003
Dick E. Hammond, Retired Professor Emeritus
College of Education
Texas State University

CONTENTS

The concept of entropy was first introduced with reference to thermodynamic phenomena and their systematic explanation in the laws of thermal physics. Subsequent development and elaboration of the concept of entropy has revealed an extremely deep and powerful notion whose explanatory utility has extended far beyond the confines of the thermodynamic phenomena that initially gave rise to the idea. But it was in physics that the concept of entropy first began its role in science. In particular, those physical phenomena that result from the conjunction of mechanical and thermal agencies are the empirical sources of the concept of entropy. The interaction between mechanical and thermal phenomena provided the basic foundation for the concept of entropy, and also provided important intellectual stimulation for the further refinement of the entropy concept, both in science and engineering. There exists a systematic relation between the entropy of macroscopic objects and the measure of available microscopic information regarding them. This systematic relation has been generalized far beyond the original conditions of thermodynamic equilibrium first studied. Furthermore, the objects entering the systematic relation through which the concept of entropy is expressed need not be physical objects at all. The idea of entropy has been so generalized that it has found fruitful application in fields far removed from its birthplace. Not only in thermal physics, but in mathematics, information theory, biology, cosmology, chemistry, linguistics, communication theory, and philosophy the concept of entropy has revealed itself as a powerful analytical tool. But the historical roots of the concept are in the physics of macroscopic thermodynamic phenomena. The attempts to understand these phenomena that arose from engineering studies of heat engines in the eighteenth and nineteenth centuries provided the initial effort toward founding thermodynamics as a science. It is in these engineering studies that one may first discern the emergence of the concept of entropy in physics.

Michael J. Harrison, "Entropy Concepts in Physics." Libor Kubat and Jiri Zeman, eds., *Entropy and Information in Science and Philosophy* (New York: Elsevier Scientific Publishing Company, 1975), p. 41.

PART 1

A Historical Review of the Entropy Concept

1. *Introduction*
2. *The Scientific Revolution to World War II*
3. *From World War II until 1970*
4. *Conclusion to the Historical Review*
5. *Afterthought*
6. *The Foundations for Education*

INTRODUCTION

Over two thousand years ago the Greeks were using the word *trope*. To them the word meant *turn, change,* or *evolve*. Like so many Greek prefixes and suffixes, *trope* is used by science today. Two examples are *tropism,* which means an involuntary orientation of an organism in response to a source of stimulation, and *phototropism,* which means turning toward the light.

The prefix *en* was traditional to ancient Greece. *Energie* equaled *into work,* while *entropie* meant *into evolve*. Therefore, the "Greek *entropie* formed an analogy of Greek *energie*."[1]

There is no evidence that a special meaning came to be attached to entropy during the decline of the Roman Empire or the following Dark Ages. Scholars simply read Greek and Latin, and change was slow.

THE SCIENTIFIC REVOLUTION TO WORLD WAR II

The work of Copernicus ushered in a new era for astronomy, for science, and, indeed, for all of humanity. Soon after the beginning of the scientific revolution a new way of examining natural phenomena began to proliferate. Physics, too, was born anew, and the scientific method attracted such men as Fahrenheit (1686-1736).

Entropy was to become associated with temperature and heat considerations. Fahrenheit developed the first reproducible and accurate thermometer. Thus, he "...made possible a systematic quantitative study of temperature and heat."[2]

By immersing hot solids into colder water, Joseph Black (1728-1799) did basic experiments which showed the transfer of heat from solids to liquids. Black's other epoch-making experiments may have lead to technological developments on the steam engine—though Watt denied this.[3]

Henry Cavendish (1731-1810) was another physical scientist from the British Isles who was interested in heat (or cold) transfers. He pondered whether melting ice generated cold, and he gave credit to Newton for considering heat as a form of motion. Cavendish's considerations, as well as those before him, pointed up the confusions surrounding the nature of heat, the difference between heat and temperature, and the direction of heat (or cold) flow.

Some of the confusion about heat was laid to rest by Count Rumford (Benjamin Thompson, 1753-1814). His experiments were well designed, important, and worth noting. Rumford had holes bored in metal cylinders to make brass cannon barrels. Horses turned a mechanism that did the boring. The cylinders always got hot, and this led him to wonder about the nature of heat. Working at a military arsenal in Munich, Rumford surrounded a brass barrel with a water-tight, wooden box. Here is the report of his findings:

> The box was filled with cold water (viz. at the temperature of 60 degrees) and the machine was put in motion.
>
> The result of this beautiful experiment was very striking, and the pleasure it afforded me amply repaid me for all the trouble I had had, in contriving and arranging the complicated machinery used in making it.
>
> The cylinder, revolving at the rate of 32 times in a minute, had been in motion but a short time, when I perceived, by putting my hand into the water, and touching the outside of the cylinder, that heat was generated; and it was not long before the water which surrounded the cylinder began to be sensibly warm.
>
> At the end of 2 hours, reckoning from the beginning of the experiment, the temperature of the water was found to be raised to 178°.
>
> At 2 hours 20 minutes it was 200 degrees; and at 2 hours 30 minutes it ACTUALLY BOILED!
>
> It would be difficult to describe the surprise and astonishment expressed in the countenances of the by-standers, on seeing so large a quantity of cold water heated, and actually made to boil, without any fire....
>
> As the machinery used in this experiment could easily be carried around by the force of one horse, (though to render the work lighter, two horses were actually employed in doing it,) these computations show further how large a quantity of heat might be produced, by proper mechanical contrivance, merely by the strength of a horse without either fire, light, combustion, or chemical decomposition; and, in a case of necessity, the heat thus produced might be used in cooking victuals.
>
> And, in reasoning on this subject, we must not forget to consider that most remarkable circumstance, that the source of the heat generated by friction, in these experiments, appeared evidently to be *inexhaustible.*
>
> It is hardly necessary to add, that any thing which any *insulated* body, or system of bodies, can continue to furnish *without limitation,* cannot possibly be a *material substance*: and it appears to me to be extremely difficult, if not quite impossible, to form any distinct idea of any thing, capable of being excited, and communicated, in the manner the heat was excited and communicated in these experiments, except it be MOTION.[4]

Still, a precise mathematical relationship between heat and work was not developed by Rumford with his ingenious experiments. The laws of thermodynamics remained to be discovered. However, by the end of the eighteenth century, calorimetry had evolved into a quantitative science.

After the turn of the century, a young Frenchman worked with, and wrote feverishly about, steam engines. Nicholas Leonhard Sadi Carnot (1796-1832) realized that his work and words applied to any kind of heat engine. By the age of twenty-eight (1824) he had published his now classic work, *Reflections on the Motive Power of Heat.*

Carnot died young, but his contributions were milestones along the road to a human understanding of energy. Among his many concepts, he made it clear that caloric (heat) always transferred from a hotter to a colder body. He stated, "Everywhere where there is a difference of temperature, and where the re-establishment of equilibrium of the caloric (heat) can be effected, the production of motive power (work) is possible."[5]

Soon after Carnot made his important contributions toward the development of the second law of thermodynamics, a German physician and physicist, Julius Robert Mayer (1814-1878) worked out a numerical estimate of the conversion factor between heat and work. At the same time, in England, the famous physicist James Prescott Joule (1818-1889) was experimenting with heat, force, and friction. Among other heat producers, he also worked with narrow water tubes, horsepower, spermoil, and an electromagnetic machine.

Joule presented his findings to the Royal Society in his classic paper, *The Mechanical Equivalent of Heat*. He and Mayer are now recognized as independently discovering the first law of thermodynamics.

No historical review of the rise of human awareness about heat and motion, the other forms of energy and entropy, would be complete without mentioning Herman Ludwig Ferdinand von Helmholtz (1821-1894). He was a great German physiologist and physicist who contributed energy formulas. He also studied another form of energy—light, as perceived by the human eye.

By the middle of the nineteenth century the basic knowledge about energy flow was being grasped by more people. In the first quarter of the century Carnot had written that:

> Since any re-establishment of equilibrium in the caloric can be used to produce motive power, any re-establishment of equilibrium which is effected without producing motive power should be considered as a veritable loss:...[6]

Also, by mid-century many scientists were aware of the foundations for the second law of thermodynamics. The knowledge was inherent in the work and words of young Carnot. As Dugdale said:

> The point here is that, given a temperature difference, this may either be used to produce work or it may be wastefully dissipated simply by a flow of heat from a high temperature to a low temperature without any work being produced. As we all know, this spontaneous flow from a hot body to a colder one will always take place if it can. This means that a situation which might produce work can be irretrievably lost by natural heat flow.[7]

Finally, the accumulated evidence proved that heat (energy) flows from an area of higher temperature (greater concentration) to an area of lower temperature (less concentration). It also became known that the process was irreversible. The heat could never flow back into the higher temperature area without the expenditure of work on the system from an outside source. Thus it was found that it was impossible to have a perpetual motion machine, and impossible to restore any system to its initial situation without using additional work (energy).

Understanding this process meant understanding entropy. When a total system is investigated through time, from its initial to its final state (equilibrium), experiments show that the heat becomes randomly dispersed. There is an irreversible increase in its entropy.

Rudolf Julius Emmanuel Clausius (1822-1888) introduced the concept of entropy in 1854. He applied the name 'entropy' to the concept in his famous paper of 1865. He was a professor of physics in Germany and Switzerland. In his paper *On The Motive Power Of Heat, And On The Laws Which Can Be Deduced From It For The Theory Of Heat,* he states:

> Carnot assumed, as has already been mentioned, that the equivalent of the work done by heat is found in the mere transfer of heat from a hotter to a colder body, while the quantity of heat remains undiminished.[8]

"The concept of entropy, which Clausius introduced and developed, is the most important single contribution made by him to science."[9]

Clausius considered the entropy of a number of systems, and his vision was very wide. He remains highly respected throughout the world of science. His statements are as true today as they were over one hundred years ago:

> For the present I will limit myself to quoting the following result: if we imagine the same quantity, which in the case of a single body I have called its entropy, formed in a consistent manner for the whole universe (taking into account all the conditions), and if at the same time we use the other notion, energy, with its simpler meaning, we can formulate the fundamental laws of the universe corresponding to the two laws of the mechanical theory of heat in the following simple form:
>
> 1. The energy of the universe is constant.
> 2. The entropy of the universe tends to a maximum.[10]

Entropy, random molecular motion, temperature, and heat became inextricably related. The first two laws of thermodynamics were to be joined by a third law when Lord Kelvin (William Thomson 1824-1907) reasoned about what kind of a temperature scale could be developed which would be zero at minimum molecular motion.

Kelvin was a Scottish physicist, mathematician, and electrician. He was not only highly capable in theory and mathematics, but his genius for invention, design, and manufacture of electrical measuring instruments set him a world apart from his contemporary physicists.[11, 12]

Lord Kelvin's reasoning about heat energy lead to his absolute zero temperature scale in 1848. 0 degrees K came to mean minimum molecular energy, zero temperature, and zero entropy. Statements like, "Temperatures lower than absolute zero are impossible," and "You cannot remove any heat from an object at absolute zero temperature," stem from Kelvin's reasoning about heat energy. 0 degrees K came to have a "...universal physical significance."[13, 14]

Kelvin's contributions greatly increased human understanding of energy flow. For with his great versatility, he was able to "...deal...with the application of thermodynamics to electrical phenomena."[15]

Slowly the forms of energy were being collected under the umbrella of thermodynamics and the entropy concept.

Another great Scottish physicist was James Clerk Maxwell (1831-1879). Working with Faraday's ideas, Maxwell produced the theory of electromagnetism. His work led to an understanding of light propagation and eventually to the creation of radio broadcasting. Perhaps "...a more important result of Maxwell's theoretical invention was the Einstein theory of relativity, with its momentous consequences for both the science and technology of our time."[16]

However, Maxwell did not confine his work to electromagnetism. He was also very concerned with the molecular motion of gases and the second law of thermodynamics. His fertile imagination led to decades of new thoughts on entropy considerations in information theory, biological evolution, and human involvement in the ordering versus disordering process in nature.

Maxwell knew that gas molecules normally diffuse and distribute themselves randomly within any system, and by so doing, this chaotic dispersion of matter confirms the second law and leads to an increase in entropy. To overcome the tendency to disorder, and thus to reduce a system's entropy, Maxwell created his now famous hypothetical "being," which remains one of the best teaching tools on the subject, and "...which could direct the course of molecules and so might cause a reaction to go in the direction of less entropy."[17]

Because of the importance of Maxwell's initial questioning, and his influence on the past one hundred years' developments in the entropy concept, his original words are best inserted here:

> A different aspect of fluctuations was considered by Maxwell in relation to the second law of thermodynamics. Imagine a gas in a vessel at a uniform temperature. 'Now let us suppose', says Maxwell, 'that such a vessel is divided into two portions, A and B, by a division in which there is a small hole, and that a being, who can see the individual molecules, opens and closes this hole, so as to allow only the swifter molecules to pass from A to B, and only the slower ones to pass

> from B to A. He will thus, without expenditure of work, raise the temperature of B and lower that of A, in contradiction to the second law of thermodynamics.' This 'small but lively' being which was invented by Maxwell was called by Thomson (Lord Kelvin) a 'demon' and is nowadays referred to as Maxwell's demon.[18, 19]

For decades Maxwell's demon was considered little more than toying with an idea—a human mind creating another paradox. There were no mathematical calculations capable of taking the demon's energy input into account, and information theory was yet to be born.

While a reasonable explanation for Maxwell's paradox was over fifty years in the future, the entropy concept itself did not stand still. The field of statistical mechanics was initiated by Maxwell, Josiah Willard Gibbs (1839-1903), and Ludwig Boltzmann (1844-1906).

Gibbs was the first American to become prominent in the evolution of the entropy concept. He was a physicist, mathematician, and chemist. There is no question that Gibbs is rightly revered for his role in the Western World's rise to rapid technological advances.

> In the hands of Willard Gibbs, thermodynamics became a powerful tool for the study of chemical reactions. His famous memoir *On the Equilibrium of Heterogeneous Substances* (1875-1878) formed a large part of the basis of physical chemistry and thus led to the establishment on a firm scientific basis of what has come to be called chemical engineering.[20]

Gibbs had a high regard for the thermodynamic concept when used by the human intellect in problem solving. During December 1901, from Yale University, Gibbs wrote in the preface to his most renowned work:

> We may therefore confidently believe that nothing will more conduce to the clear apprehension of the relation of thermodynamics to rational mechanics, and to the interpretation of observed phenomena with reference to their evidence respecting the molecular constitution of bodies, than the study of the fundamental notions and principles of that department of mechanics to which thermodynamics is especially related.[21]

Boltzmann's mathematical work on entropy anchored the idea of order to statistical probability and paved the way for further development of the entropy concept. He was a great Austrian theoretical physicist, so great that some authors regard him alone as the founder of statistical mechanics.

> It was, perhaps, natural that people should try to link the two points of view, i.e. the thermodynamic and the mechanical. Boltzmann

> (1844-1906) was the first to succeed in interpreting entropy from a microscopic point of view. The kinetic theory of gases was well known to him; he developed it and produced from it a most powerful method of studying the thermodynamic properties of large numbers of similar molecules. He was in fact the founder of what we now call 'statistical mechanics'.[22]

Boltzmann apparently saw contradiction and paradox in his own work. These were problems which could only be solved with the rise of quantum theory.

> But in Boltzmann's day the quantum theory had not been formulated ... Towards the end of his life Boltzmann became pessimistic and severely depressed, torn, no doubt, between a conviction that he was right and an inability to prove it. In 1906 he committed suicide. It is ironical that by this time his great contributions to physical theory had been largely vindicated and had provided the foundation on which Planck built his theory of quanta. It was this theory which ultimately made it possible to resolve many of the paradoxes that had plagued Boltzmann's own work.[23]

However, Boltzmann's generalization of the concept of entropy was unsurpassed to his time. For he was able to understand and to explain the entropy of a gas even when it was not at equilibrium.

> This conceptual generalization provided Boltzmann with the vehicle to release statistical mechanics from its initial connection to the kinetic theory of gases and the more restricted methods of that latter discipline .... Later, Boltzmann's research was developed further by Planck who extracted the statistical implications of Boltzmann's thought and summarized his definition of entropy S in a statistical relation equivalent to that given by Boltzmann:
>
> $$S = k \log W$$
>
> From this formula the first and second laws of thermodynamics are readily derived.[24]

Because of the importance of this formula, it is further explained in Appendix C. Indeed, the formula was deemed so important a contribution to the human understanding of natural phenomena that it "...was placed on Boltzmann's tombstone."[25]

> Boltzmann may be regarded as the first to have appreciated the role of statistical considerations in the concept of entropy. The

> expression S = k log W appears on Boltzmann's grave in the Central Cemetery of Vienna. His formula represents his great gift to our understanding of the physical universe and the observed directions of temporal change exhibited by natural processes wherever and whenever they occur.[26]

By the beginning of the twentieth century, the work of Maxwell, Gibbs and Boltzmann had been completed, and the entropy concept was firmly stabilized in physics, chemistry and engineering. Classical thermodynamics had already been molded and new directions initiated. For closed, isolated, and equilibrium-bound systems, humans had made progress by explaining natural phenomena in a new way. This made further advances in the technology of the twentieth century possible. But Maxwell's paradox was still left unexplained, and the door remained open for further developments in entropy considerations of non-equilibrium systems. As so often happens in human inquiry, solving one problem also initiates many new questions. After recording Boltzmann's great contribution to thermodynamics, Lindsay follows with these words:

> The statistical interpretation of entropy and the principles of thermodynamics leads at once to another important point of view, namely, that of the role of *order* in the direction of naturally occurring processes. It does not take much reflection to see a profound connection between statistical probability and order .... We can therefore introduce another interpretation of entropy. Increase in entropy means a transition from a more orderly state to a less orderly state.... On this interpretation, the meaning of the second law of thermodynamics, the law of increasing entropy, is now clear. In any naturally occurring process the tendency is for all systems to proceed from order to disorder, and the maximum entropy of Clausius is the state of complete disorder or thorough randomness, out of which no return to order is practically possible because it applies to the universe as a whole and nothing short of an inexpressibly improbable revolution could reverse the process and decrease the entropy. From this point of view, the trend from order to disorder with production of entropy is inexorable. The second law always wins in the end. A gloomy outlook indeed! But there is perhaps a silver lining in the cloud.[27]

From at least the time of Clausius onward there was seen to be an impending doomsday built into the universe. The second law and the entropy concept seemed to indicate that the universe was running down or going down hill. Eventually all the matter and energy would be in equilibrium, randomly distributed, and thus a complete state of disorder would be reached for any system, including that largest of all known systems—the universe.

Apparently, at the beginning of the twentieth century, there was agreement among the many European physical scientists about the power and importance of the entropy concept. But relatively few scientists spent time working with entropy beyond its technological applicability. More discoveries and contributions were immediately available by specializing in electricity and electronics, and especially in elementary particle and nuclear physics and chemistry.

There was an obscure little article by R. Emden in a 1938 *Nature* magazine. Fast wrote of it in the introduction to his book, *Entropy*:

> With the help of examples from widely divergent fields of physics, chemistry and technology we shall show further that the concept of entropy has a significance no less fundamental than that of energy. On the relation between energy and entropy R. Emden writes: "As a student, I read with advantage a small book by F. Wald entitled *The Mistress of the World and Her Shadow.* These meant energy and entropy. In the course of advancing knowledge the two seem to me to have exchanged places. In the huge manufactory of natural processes, the principle of entropy occupies the position of manager, for it dictates the manner and method of the whole business, whilst the principle of energy merely does the bookkeeping, balancing credits and debits."[28]

Still, while the entropy concept was deemed important, it seemed to stagnate in lieu of the many other needs of advancing technology in the Western World. What would eventually place entropy at the focal point of consideration again were the ideas from another branch of science—biology.

Early in the twentieth century Darwin's theory of biological evolution had already caused fifty years of rumbling, and Mendel had experimented with the mechanism whereby evolution could function. Finally it came to be realized that all living things use energy and matter to grow and go. And in so doing, for a time, their systems become less randomized and more organized. With the rise of experimental embryology and developmental anatomy, vitalism finally lost its vogue in biology.[29]

Maxwell's demon could be looked at in a new light—an intelligent being tampering with the normal flow in a system. "For years there seemed to be no reasonable explanation for this paradox. However, Szilard pointed out in 1929 that Maxwell's demon requires the use of information."[30]

Leo Szilard (1898-1964) published the explanation for Maxwell's paradox in a German physics journal of 1929, (*Zeitschrift fur Physik*, 53, 840-856). The now famous article was not reprinted in English until 1964, (*Behavioral Science*, Vol. 9. No. 4, October). The translated title of the article is, *On The Decrease Of Entropy In A Thermodynamic System By The Intervention Of Intelligent Beings.*

> This is one of the earliest, if not the earliest paper, in which the relations of physical entropy to information (in the sense of modern mathematical theory of communication) were rigorously demonstrated

> and in which Maxwell's famous demon was successfully exorcised: a milestone in the integration of physical and cognitive concepts.[31]

World War II brought Szilard to the United States as it did so many great European scientific minds. His interest in atomic energy pulled him together with Einstein, Fermi, and other important members of the Manhattan Project.

FROM WORLD WAR II UNTIL 1970

After World War II Szilard's interest remained with radiation considerations, and like so many atomic scientists, after the War he became politically involved because of an understanding of radiation's detrimental effects. He worked in biophysics and biochemistry, and left the entropy concept to evolve in the writings of others.

Thus, the first half of the twentieth century gave its greatest acclaims to the work being done on the nucleus of atoms, perhaps because of two world wars and the very rapid rise of sophisticated military technology. However, soon after World War II interest in the entropy concept was to swell again in Europe, and entropy considerations were to blossom in the United States as never before.

> In 1949 Claude Shannon of Bell Telephone Laboratories published his classic paper, entitled "The Mathematical Theory of Communication" (in Shannon and Weaver, 1949), which is often said to have "founded" the field of information theory...
>
> I shall extend the entropy concept primarily through the apparatus of information theory, but I shall extend that also. Shannon (1949) gave the most general definition of entropy to date, and I shall extend his concept.[32]

Claude Elwood Shannon (1916-2001) is the American applied mathematician who made the postwar attempt to link information theory with the entropy concept. Though Shannon is referred to as fathering information theory, he was not alone in his thinking. The year before his important paper appeared, the American mathematician Norbert Wiener (1894-1964) had published his most thoughtful book, *Cybernetics*. Whereas Shannon was primarily concerned with entropy and communications engineering, Wiener realized that the control over any system's behavior, be it an organism or a mechanism, involves communication.[33]

Living systems were unavoidably being considered in the information and entropy area, just as Maxwell and Szilard had foreseen. And in the same decade added impetus was given to the life and entropy considerations by Schrodinger.

Erwin Schrodinger (1887-1961) was an Austrian physicist and philosopher who is still popular for his book, *What is Life?* Schrodinger saw that in the evolution of life there was a tendency for higher and higher organization and complexity and flexibility. Unlike the ordered and rigid and low-entropy systems of the crystals, life's systems reduced entropy in a manner which allowed for change. From the lowly housefly and the termitary of the termite, Schrodinger saw life as going from disorder to order.

Schrodinger's book may be one of the earliest popular accounts in English of the relation between entropy, order, and life. He saw life as a massive movement to reduce entropy. Or, as he put it, the living organism "...feeds on negative entropy."[34]

These last two words were to be joined later by an important American theoretical physicist. Leon Brillouin (1889-1969) called the human trend to put more and more efficiency, order, and organization into their information-communication-knowledge systems the negentropy process. Whereas entropy meant going from a more ordered state to a more random state, negentropy meant going from a more disordered state to a more organized one.[35]

In the same decade of the forties, when Schrodinger, Wiener, Shannon, and others were contemplating and joining life and information to the entropy concept, two other contemporary men were in their embryonic states, of writing about thermodynamics and the entropy concept. Ilya Prigogine (1917-2003) was in Europe, already published by 1947, deeply interested in minimum entropy production, and, for his work on the subject, destined to win the Nobel Prize thirty years later. Callen says:

> Accordingly Prigogine showed that, in the steady state which is reached when certain affinities are constrained to have definite values, and in the absence of a magnetic field, all unconstrained affinities assume the values which minimize the entropy production function. Prigogine's reformulation did suggest a generalization, although his conjectured generalization does not appear to have been formulated explicitly in the literature.[36]

Callen's reference for these statements is an earlier work by Prigogine, *Etude thermodynamiques des phenomenes irreversibles*, published at Desoer, Liege, in 1947.

Robert Bruce Lindsay (1900-1985) was in the United States writing physics textbooks on statistical mechanics, thermodynamics, and other material which would eventually lead to the delineation of his "Thermodynamic Imperative". Lindsay explains how humans can have guidelines for their behavior–based on what science knows about natural phenomena, the laws of thermodynamics, and the entropy concept of human consumption of entropy.[37]

In the forties, philosophizing about entropy and living systems was way ahead of the mathematicians' capabilities, and many physicists and chemists thought therefore that they had the only right of domain over the entropy concept. Biologists, sociologists, and other scientists seemed reluctant to wed the entropy concept with the knowledge in their own fields.

The classical thermodynamics of isolated and closed systems was sufficient in university courses and industries for almost any technological advancements. So why extend the role of thermodynamics? Prigogine explains in his own words:

> A serious limitation of classical thermodynamics as a general tool for the macroscopic description of physico-chemical processes lies in the fact that its method is based on such concepts as "reversible processes" and "true equilibrium states."
>
> It is now generally recognized that in many important fields of research a state of true thermodynamic equilibrium is only attained in exceptional conditions. Experiments with radioactive tracers, for example, have shown that the nucleic acids contained in living cells continuously exchange matter with their surroundings. It is also well known that the steady flow of energy which originates in the sun and the stars prevents the atmosphere of the earth or stars from reaching a state of thermodynamic equilibrium.
>
> Obviously then, the majority of the phenomena studied in biology, meteorology, astrophysics and other subjects are irreversible processes which take place outside the equilibrium state.
>
> These few examples may serve to illustrate the urgent need for an extension of the methods of thermodynamics so as to include irreversible processes. Such a generalization is all the more important because the general microscopic theory of irreversible processes is still in its initial stages....
>
> It is useful to classify thermodynamic systems according to the exchanges of energy (heat and work) and matter through their boundaries. We shall distinguish between *isolated systems* which can exchange neither energy nor matter, *closed systems* which exchange energy but no matter and *open systems* which exchange both energy and matter with the exterior.
>
> Classical thermodynamics has been concerned mainly with the study of closed systems. One striking achievement of recent developments has been to withdraw this limitation so as to generalize the methods of thermodynamics to open systems which are of great importance for biological thermodynamics as well as for many other fields such as meteorology and geology.[38]

Prigogine was quite aware that living systems were open systems, exchanging energy and matter with their exterior. In the early fifties he wrote:

> ... the fact that during growth living organisms actually show a decrease of entropy production during evolution up to the stationary state. Also, the fact that their organization generally increases during this evolution corresponds to the decrease of entropy as studied...The behavior of living organisms has always seemed so strange from the point of view of classical thermodynamics that the applicability of thermodynamics to such systems has often been questioned. One may say that from the point of view of the thermodynamics of open and stationary systems a much better understanding of their principal features is obtained.[39]

With funding from the United States Air Force, General Motors Corporation, General Electric, and others, Prigogine was able to continue his mathematical work with open systems. He was also supported by the Free University of Brussels, the University of Chicago, Northwestern University, and the University of Texas, where he was the director of the Center for Statistical Mechanics and Thermodynamics.

Prigogine's quantitative analysis, comments, acknowledgements, and extensive bibliography can be found in his work on non-equilibrium. Published in 1962, the bibliography reveals the massive amount of work Prigogine had to review from World War II until that time.[40]

Lindsay's completed work on the non-mathematical treatment of the total entropy concept also appeared in the early sixties. *The Role of Science in Civilization* was first published by Harper and Row in 1963. But Lindsay's idea of entropy reduction and his phrase "The Thermodynamic Imperative" are found in his earlier article, "Entropy Consumption and Values in Physical Science".[41]

At the time the article was published, Lindsay was the Dean of the Graduate School and Professor of Physics at Brown University. The article has no bibliography, but a follow up article by Steven Polgar a the University of California at Berkeley has an extensive bibliography which reveals the amount of non-mathematical writing that was done between World War II and the sixties.[42]

Polgar's literature list stretches from Johnstone's book (*The Mechanism of Life in Relation to Modern Physical Theory,* London: Longmans, 1921) to Lindsay's article mentioned above. What makes Polgar's article unique, and often quoted, is the fact that it consolidated the thoughts of so many great men in biology, sociology, anthropology, and evolution.

Polgar said that *persistence* (the perpetuating mechanisms), *replication* (chromosomal autocatalysis), and *environmental modification* were entropy reducing, or as he called it, "entropy retarding".

> Whatever the total effect of entropy, modification of the environment by a living organism usually decreases the chances for

> "disorder" from its local point of view by increasing its chances for survival as a species.[43]

He also stated a fourth cause for "entropy retardation" in living systems. "Three processes of entropy retardation have been delineated, *structuring,* it seems, must be a fourth."[44]

Polgar expanded on his idea of structuring:

> The building of settled villages meant a more intensive modification of the environment, eventually leading to the deliberate nurturing of animals and plants. This constituted purposeful replication of available energy for future use . . . . Until now life has succeeded in maintaining itself: its persistence has retarded entropy. It has persisted by generating order from order, i.e. replication, and by modifying its environment. Replication, furthermore, has proceeded in the direction of more and more structuring, and this structuring has accompanied an increasingly wider integration of systems until today the whole planet, in many respects, is just one single system. Could the recognition of this trend constitute yet another way of retarding entropy?[45]

Whereas Lindsay saw the ethical principle in terms of the human *necessity* to consume as much entropy as possible, Polgar's ethical statement includes *recognition* of the entropy reduction obligation.

> The thermodynamic imperative itself may thus be said to have undergone evolution, from the unconscious adaptations of infrahuman organisms, through the development of "exosomatic" culture (Lotka, 1945) by which stores of information can be transmitted to subsequent generations and otherwise "distributed" (Brillouin, 1950), to the scientific descriptions of evolution and the deliberate attempts to increase the availability of energy for survival and the enjoyment of life. Given the insignificance of life on this planet when viewed from astronomical perspectives, these considerations lead to the following ethical statement: While the universe will probably continue to exist and to change, the likelihood of life-somehow-related-to-life-as-we-know-it surviving for a long time is small, but the chances can be improved by recognition of the thermodynamic imperative and action based upon it.[46]

Carried one step further, if recognition of entropy reduction is important, then learning and teaching about the entropy concept becomes an important experience. Education in entropy reduction might become an entropy reducing enterprise in itself.

CONCLUSION TO THE HISTORICAL REVIEW

Many contributors, books, and articles have not been mentioned in this historical review of the entropy concept. However, most of the material mentioned here contains adequate bibliographies. Because of the proliferation of the entropy concept, this conclusion attempts to concentrate on the writings since 1970. The books and articles below contain further up-to-date references.

Blum's *Time's Arrow and Evolution* was reprinted for the sixth time in 1970. It contains a new chapter on "Order, Negentropy, and Evolution."[47]

In 1971, Prigogine was published again with a member of the General Motors Research Laboratories on the *Kinetic Theory of Vehicular Traffic*. The book applies his open-systems, non-equilibrium, irreversible statistics to a distinctly human activity.[48]

Also in 1971, Bent's article was published on "Haste Makes Waste—Pollution and Entropy." His article ties maintenance of the environment to the entropy concept.[49]

In 1972 Gatlin's book on *Information Theory and the Living System* was published, with many examples of entropy reduction via information storage by living systems.

> Let us take a simple example. A library obviously contains stored information. The information is stored in a linear sequence of symbols ordered according to the constraints of a language. The sequences are organized into books and periodicals, and these are carefully ordered on shelves and neatly catalogued. Everywhere order and constraint are associated with the information storage process. This is a state of lowered entropy. If we were to take each page of each book, cut it into single-letter pieces and mix them in one jumbled heap, the entropy would unquestionably increase, but the stored information would decrease. Stored information, the "what we do say" of Weaver's statement, is associated with the ordering process brought about by the constraints of a language or any organized information storage process. Since stored information varies inversely with entropy, lowered entropy means a higher capacity to store information.[50]

Prigogine and others published again in 1972. Their two-part article on the "Thermodynamics of Evolution" takes into account Prigogine's work on *open systems, entropy flow, non-equilibrium order*, and the rise of knowledge about the DNA code. Their subtitle is "The functional order maintained within living systems seems to defy the Second Law; non-equilibrium thermodynamics describes how such systems come to terms with entropy." The authors show that through the organization of macromolecules which are not static, "this primitive memory effect makes these structures capable of storing information accumulated in a remote past."[51]

And they show further "...that nucleic acids are a necessary prerequisite for self-organization because they possess the ability to act as templates."[52]

In 1973, ten years after its original printing, Lindsay's *The Role of Science in Civilization* was reprinted. His ethical code remains the same:

> Considerations of this kind seem to suggest a new kind of imperative which if reasonably interpreted might serve as a satisfactory basis for an ethical code. We shall call it the *thermodynamic imperative* and phrase it as follows: All men should fight always as vigorously as possible to increase the degree of order in their environment, i.e., consume as much entropy as possible, in order to combat the natural tendency for entropy to increase and for order in the universe to be transformed into disorder, in accordance with the second law of thermodynamics.[53]

Also, in 1973, Lehninger's book *Bioenergetics* was printed for the second time. His final paragraph suggests the human brain as an entropy reducing mechanism:

> Perhaps the most challenging of all problems in bioenergetics is posed by the phenomenon of memory, particularly the highly developed ability of man to store relatively immense amounts of information in his memory. We have already seen that information has a thermodynamic dimension, since it is the opposite of entropy. Thus there must be a finite energy cost wherever information is laid down in memory. How memory is stored in the human brain, whether in the form of specific interneuronal connections or in the form of macromolecular structure, such as is used for storage of genetic information, is still totally unknown. Perhaps the molecular basis of memory, thought, and behavior will be among the ultimate biological secrets to be solved by man, secrets that will come close to defining man's own being. Thus we see that in its broadest aspects bioenergetics spans some of the most remarkable and specialized activities in the world of living organisms.[54]

Although the *Project Physics Reader #3* had mentioned entropy considerations as early as 1970, it was between 1973 and 1977 that the entropy concept was placed in high school and college freshman science textbooks. But the concept was only on a very elementary level, rarely mentioning biological or human behavioral considerations. Saperstein's college freshman text *Physics: Energy in the Environment* of 1975 was the first to expand the entropy concept with many new ideas and examples. In his section on "Entropy as a Measure of Randomness" he explains how "randomness increases with uncertainty," "entropy in history," how "entropy can decrease in non-isolated systems," "entropy in agriculture," and finally in "entropy, information, and noise," how "noise is entropy":

> Since information is the lack of entropy, putting information into a system is the same as removing entropy from the system. But we know the consequences of decreasing the entropy of a physical system. Thus the gaining of information, or the increase in our knowledge, requires an input of energy and the increase in the entropy of the rest of the universe. Information does not come free of the expenditures of energy.
>
> When trying to obtain information (receive a "message"), we are always confronted with the problem of noise: static in a radio receiver or TV set, random sound or other peoples' voices when trying to engage in conversation, distractions when trying to read. Noise may be defined as energy (e.g., sound energy) with no discernible pattern. It interferes with, and may overwhelm and hide, the information-containing pattern we are trying to receive. Hence noise hinders information flow, and increases entropy. Thus, noise is a form of entropy just as information is a form of "lack of entropy" or *negentropy*.[55]

In 1975 the above mentioned people were not alone in their writings and considerations about the entropy concept. Kubat and Zeman (editors) published their compiled articles in 1975. The *Entropy and Information in Science and Philosophy* volume contains eighteen articles on entropy concepts, ranging from physics through all the disciplines to information theory and philosophy. The collected papers bring together some of the great minds of both Eastern and Western nations.[56]

In 1976 Fox wrote an article entitled "The Recolored Mentality—Ethical Lessons From Science." He states that "The Second Law of Thermodynamics also poses an interesting dilemma for students. Man's physical body and his social aspirations are anti-entropic, representing states of higher organization and concentration of potential energy in a sense." He further proposes that "...students build hope in something ethically durable." Fox's purpose in writing the article was to show that "there are ethical lessons arising from conceptual science..." and that "these lessons may properly be dealt with in secondary school science."[57]

By March of 1977 the Georgia Conservancy reprinted its *Wolfcreek Statement* and the United States was on its way to having a new Energy Department. An effort would be made on a national level to slow down the irretrievable loss of energy.[58]

Slowing the speed at which energy, matter, or information is lost results in reducing entropy. But Bent did not like the government's use of the term "energy conservation." In his article of May 1977 he says, "SavEnergy, we are told. There's an energy crisis. But energy is always saved. By definition . . . . There is no need to try to conserve energy. Energy is always conserved." Since experimental evidence shows that energy is neither created nor destroyed, Bent was taking issue with the government's term and the national energy policy, and not with the entropy concept:

> What Washington is asking us to do in the present "energy crisis" is to save not energy but *transformable* energy—although not exactly that either. For a gas at high pressure, for example, has no more energy pressure. What it has is less entropy and, thereby, a greater capacity—in passing spontaneously to higher-entropy, less-ordered states—to do useful work.
>
> What we are being asked to do, it appears, is to *husband spontaneous events*. Live leanly. Avoid producing entropy unnecessarily. *Blessed are the low-entropy producers, for they shall inherit the Earth,* is not a value judgement. It is a prediction.
>
> What we don't need is a national energy policy. What we do need is a *personal entropy ethic.* Its achievement, however may require several winters of discontent ....
>
> The Second Law of Thermodynamics does not tell us, however, how fast entropy can or will or should increase. In this century, as in the last, the energy crisis will probably do more for thermodynamics than thermodynamics will do for the energy crisis.[59]

By the fall of 1977 Bent was on the road, holding workshops for the Office of Science Education of the American Association for the Advancement of Science (AAAS), with financial support from the National Science Foundation (NSF). His Thermodynamics, Art, Poetry, and the Environment workshops continued into the spring of 1978, and they covered:

> ... development with students of (1) a scenario for a film that, projected backwards, dramatizes time's arrow; (2) a script for a play on the theme The Entropy Ethic; (3) a critique of the commandment, Thou shalt not unnecessarily create entropy; (4) analogies between science and the arts; (5) statistical data (hand- or computer-generated) that illustrate Boltzmann's Energy Distribution Law.[60]

Also, in the fall of 1977, Stein added to Polgar's dimension by tying energy use to building design in his book *Architecture and Energy*. Watson, of Yale's School of Architecture, reviewed the book:

> These issues will increasingly be brought into the economic and political realm. We are able to see to the end of our existing fossil-fuel base, so the heating and cooling plant of existing buildings fueled by gas or oil will probably have to be replaced, with the attendant conversion costs, well within the useful life of the building. Beyond this, energy use is an ethical and social issue: "Buildings designed to overuse energy

penalize everyone who needs energy for survival," Stein writes in showing the connection between energy use in the United States—35 percent of all energy use globally—and subsistence living conditions in much of the developing world.[61]

Thus, just as Polgar had suggested, human structuring was brought under the umbrella of thermodynamics and ethical ideas, along with many other behavioral considerations.

As 1977 drew to a close, *Science* magazine reviewed "The 1977 Nobel Prize in Chemistry":

> The departure from classical equilibrium thermodynamics started with two famous papers of L. Onsager published in 1931, for which he was awarded the 1968 Nobel Prize in Chemistry ....
>
> This important result, known as Onsager's reciprocal relations, opened the way for an extensive discussion and application of the thermodynamics of near-equilibrium phenomena ....
>
> Prigogine proved this implication with great generality in 1945 and called it the principle of minimum entropy production . . . .
>
> The dissipative structures can be maintained only through a sufficient flow of energy and matter. The work required to maintain the system far from equilibrium is the source of the formation of order . . . .
>
> The possibility of building order through fluctuations under extreme non-equilibrium conditions has implications for biomorphology, embryology, and evolution. The maintenance of highly ordered structures requires exchange of energy and matter with the surroundings, and only this constant flux supports the high degree of organization needed for existence.[62]

On January 1, 1978 Prigogine spoke about his Nobel Prize and the concept of minimum entropy production. He closed by stating that he saw the concepts unraveling as being "the wave of the future."[63]

As 1978 began, newspapers and popular magazines had dozens of articles reviewing the 1977 crimes of violence, vandalism, and other destructive behaviors. Also in January of 1978 *Phi Delta Kappan* devoted its entire special issue to "The Problems of Discipline and Violence in American Education."[64]

The plea became more insistent from many quarters of society for education to help give young students better ethical instruction.

AFTERTHOUGHT

This historical review of the entropy concept has moved from the human concern about the flow of hot or cold to deliberations about an entropy ethic. It has traversed the road from the invention of the thermometer to statistical physics for open systems. Along the way, the disciplines of classical thermodynamics, evolution, developmental anatomy, genetics, information theory, and bioenergetics were developed, to mention only a few of the newer courses. Many of the important drivers during this three-hundred-year trip through time are listed chronologically in Appendix D.

Unfortunately, many of the other important developments and discoveries by science were passed by. It must be understood that the entropy concept was pushed along by knowledge about the periodic table of the elements, chemical bonding, and the sources and forms of energy. These monuments to the human brain have only been touched on here: however, their contributions were necessary for the continuing evolution of the entropy concept. Also, entropy concepts in the theory of music and the other arts have not been treated here. And, unmentioned, are the philosophical aspects of economics, religions, sports, governments, and other human institutions. References to the impact of the entropy concept on these other branches of knowledge are readily available in university libraries, and many references can be found by reading the material already mentioned, as well as their bibliographies.

The history of the entropy concept shows an unraveling of a new foundation for ethics based on empirical knowledge about natural phenomena.

A final tribute is due to the thoughtfulness of Albert Einstein. He seemed to sense the possibility of using scientific findings as guidelines for human conduct when he said, ". . . ethical directives can be made rational and coherent by logical thinking and empirical knowledge."[65]

THE FOUNDATIONS FOR EDUCATION

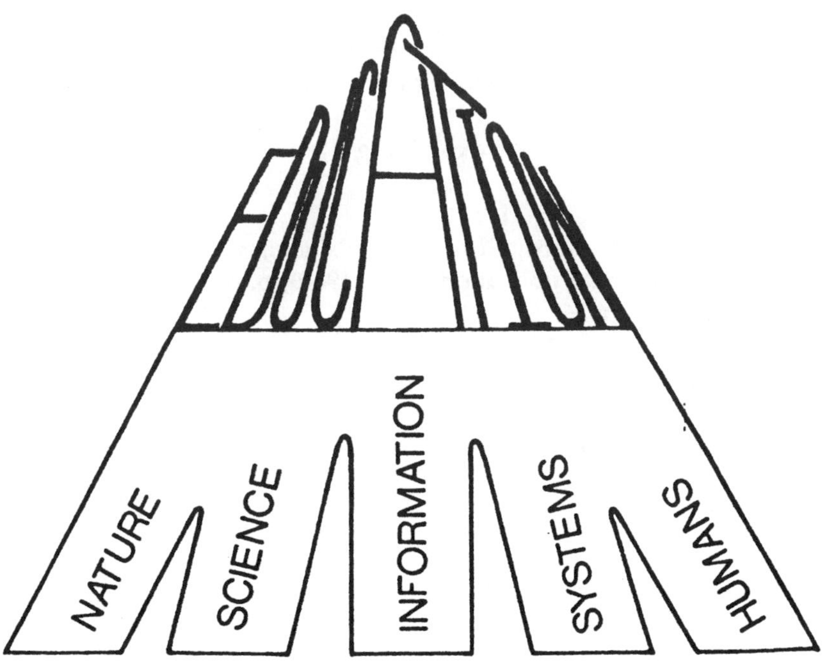

---

[1]Ernest Klein, *A Comprehensive Etymological Dictionary of the English Language* (New York: Elsevier Publishing Company, 1971), p. 250.

[2]J.S. Dugdale, *Entropy and Low Temperature Physics* (London: Hutchinson & Company, 1966), p. 13.

[3]Robert Bruce Lindsay. *The Role of Science in Civilization* (Westport: Greenwood Press, 1973), p. 210.

[4]Dugdale, pp. 16-17.

[5]J.S. Ames, ed., *Harper's Scientific Memoirs VI: The Second Law of Thermodynamics* (New York: American Book Company, 1899), p. 8. In this memoir edition the works of Carnot, Clausius and Thomson are translated by W.F. Magie.

[6]Ames, p. 13.

[7]Dugdale, pp. 36-37.

[8]Ames, p. 88.

[9]Ames, p. 108.

[10]Dugdale, pp. 71-72.

[11]Lindsay, pp. 214-215.

[12]Ames, pp. 147-148.

[13]Alvin M. Saperstein, *Physics: Energy in the Environment* (Boston: Little, Brown and Company, 1975), pp. 128, 232-233.

[14]Dugdale, p. 43.

[15]Ames, p. 147.

[16]Lindsay, p. 213.

[17]Harold F. Blum, *Time's Arrow and Evolution* (Princeton: Princeton University Press, 1970). p. 109.

[18]Dugdale, p. 152. In this quote, Maxwell's words are taken directly from his book, *Theory of Heat* (London: Longman, Green and Co., 1883), p. 328.

[19]A diagram, further explaining Maxwell's words, is contained in Appendix B.

[20]Lindsay, pp. 213-214.

[21]J. Willard Gibbs, *Elementary Principles in Statistical Mechanics* (New York: Dover Publications, Inc., 1960), p. ix.

[22]Dugdale, p. 76.

[23]Dugdale, p. 137.

[24]Michael J. Harrison, "Entropy Concepts in Physics," Libor Kubat and Jiri Zeman, eds., *Entropy and Information in Science and Philosophy* (New York: Elsevier Scientific Publishing Company, 1975), p. 46.

[25]Lindsay, p. 161.

[26]Harrison, p. 47.

[27]Lindsay, pp. 161-162.

[28]J.D. Fast, *Entropy* (Eindhoven, Holland: N. V. Phillips, 1968), p.4.

[29]Joseph Needham, *Order and Life* (New Haven: Yale University Press, 1936).

[30]Albert L. Lehninger, *Bioenergetics* (Menlo Park, California: W.A. Benjamin, Inc., 1973), p. 181.

[31]Bernard T. Feld and Gertrud Weiss Szilard, eds., *The Collected Works of Leo Szilard: Scientific Papers* (Cambridge, Massachusetts: The MIT Press, 1972), p. 120.

[32]Lila L. Gatlin, *Information Theory and the Living System* (New York: Columbia University Press, 1972), pp. 20-22.

[33]N. Wiener, *Cybernetics* (New York: John Wiley and Sons, 1948).

[34]Erwin Schrodinger, *What Is Life?* (New York: Macmillan, 1945), especially Chapter 6.

[35]L. Brillouin, *Science and Information Theory* (New York: Academic Press, 1956). See, in particular, Chapter 9.

[36]Herbert B. Callen, "On The Principle of Minimum Entropy Production," I. Prigogine, editor, *Proceedings of the International Symposium on Transport Processes in Statistical Mechanics* (New York: Interscience Publishers, Inc.. 1958), pp. 327-333.

[37]Lindsay, Chapter 9.

[38]Ilya Prigogine, *Thermodynamics of Irreversible Processes* (New York: Interscience Publishers, 1961), pp. v and 3.

[39]Ilya Prigogine, *Introduction to Thermodynamics of Irreversible Processes* (New York : Interscience Publishers, a division of John Wiley & Sons, 1955), p. 92.

[40]I. Prigogine, *Non-equilibrium Statistical Mechanics* (New York: Interscience Publishers, a division of John Wiley & Sons, 1962), pp. 308-312.

[41]R. B. Lindsay, "Entropy Consumption and Values in Physical Science," *American Scientist*, (1959), pp. 376-385.

[42]Steven Polgar, "Evolution and the Thermodynamic Imperative," *Human Biology*, (1961), pp. 99-109.

[43]Polgar, p. 101.

[44]Polgar, p. 103.

[45]Polgar, pp. 105-107.

[46]Polgar, pp. 107-108.

[47]Blum, Chapter 13. Blum is at Woods Hole, Princeton, and the State University of New York at Albany.

[48]Ilya Prigogine and Robert Herman, *Kinetic Theory of Vehicular Traffic* (New York: American Elsevier Publishing Company, Inc., 1971).

[49]Henry A. Bent, "Haste Makes Waste—Pollution and Entropy," *Chemistry* (October, 1971), pp. 6-15. Bent is Professor of Physical Chemistry at North Carolina State University.

[50]Gatlin, p. 49. Gatlin is Assistant Research Biophysicist at the Space Sciences Laboratory, University of California at Berkeley.

[51]Ilya Prigogine, Gregoire Nicolis and Agnes Babloyantz, "Thermodynamics of Evolution." *Physics Today*, (November, 1972), p. 26. Prigogine still holds a dual appointment as professor of physical chemistry and theoretical physics at the Free University of Brussels and director of the Center for Statistical Mechanics and Thermodynamics at the University of Texas at Austin. Nicolis and Babloyantz are both at the Free University of Brussels.

[52]Ilya Prigogine, Gregoire Nicolis and Agnes Babloyantz, "Thermodynamics of Evolution," *Physics Today*, (December, 1972), p. 40.

[53]Lindsay, p. 292.

[54]Lehninger, p. 228.

[55]Saperstein, pp. 245-253.

[56]Libor Kubat and Jiri Zeman, eds., *Entropy and Information in Science and Philosophy* (New York: Elsevier Scientific Publishing Company, 1975).

[57]Fred W. Fox, "The Recolored Mentality—Ethical Lessons From Science," *The Science Teacher*, (September, 1976), pp. 22-24. Fox is professor of science education at Oregon State University.

[58]*The Wolfcreek Statement: Toward A Sustainable Energy Society* (Atlanta, Georgia: The Georgia Conservancy, March, 1977).

[59]Henry A. Bent, "Entropy & the Energy Crisis," *The Science Teacher*, (May, 1977). pp. 25-30.

[60]*NSF Chautaugua-Type Short Courses for College Teachers* (Washington, D.C.: Office of Science Education, American Association for the Advancement of Science, Fall, 1977).

[61]Donald Watson, "Energy Use and Building Design," *Science*, (28 October, 1977), pp. 391-392.

[62]Itamar Procaccia and John Ross, "The 1977 Nobel Prize in Chemistry," *Science* (18 November, 1977), pp. 716-717.

[63]Ilya Prigogine speaking over WOAI FM Radio Station in San Antonio, Texas at 2:30 AM on January 1, 1978.

[64] *Phi Delta Kappan,* (January, 1978).

[65] Albert Einstein, "The Laws of Science and the Laws of Ethics," *Out of My Later Years* (New York: Philosophical Library, 1950), p. 114.

PART 2

Education and the Entropy Concept

INTRODUCTION

Within that largest of all voids, the billions of galaxies move rapidly to undetermined destinations. The Milky Way spins on its gigantic axis, carrying the earth along a wave pattern, as it takes its seemingly endless journey through space. Humans move in the totality of the reality called nature. Motion is a fundamental law of the universe. Entropy is the law underlying all laws explaining motion.

It is not possible, on earth or anywhere in its environment, to create or destroy energy-matter. Energy and matter simply move from one place to another. In so doing, they become less concentrated, less arranged, less ordered, less organized, more dissipated, more dispersed, and more randomized. This is the law of entropy. This is the law of nature.

Using the same energy as the stars, and composed of the same matter as the rest of the galaxy, life on earth is not an exception to the law of entropy. Life is a part of nature. Yet, life is different from the rest of nature. Life lives on, in its millions of forms, year after year, appearing to defy the law of entropy. Life's myriad of highly organized forms have found a way to control entropy—a way to maintain their highly ordered states of existence.

The law of entropy is not overthrown, nor is it circumvented by life on earth. What life has done is protect the organization of matter which makes prolonged existence possible. The genes live on, one generation after another, wending their way through the branches of biological evolution. However, the rigors of the law of entropy take their toll. Since all living things must die, many organisms are needed to protect the arrangement of matter in the genes. Each organism uses a constant supply of energy and matter to maintain himself and to move. These needs and uses control entropy by leading to organization within a living organism, and within the organization of groups of organisms. The price is paid to the law of entropy by an increase in entropy outside the organisms.

Humans now have some knowledge about the game of managing entropy. They know some of its rules, know they must play, and they know they cannot win. However, the law of entropy does not dictate how fast a living system must lose.

Information exists in the arrangement of matter. Thus, the organization of matter and information are the same. The more organized the matter, the more information, and the more entropy is managed. As organisms protect their highly ordered genetic structures, they are protecting the information which keeps the entropy-control process intact from one generation to the next.

Humans have evolved to a state of existence different from other life forms. While all of life depends on matter and energy from the environment, and all must protect the information in their chromosomes, humans can store vast amounts of information outside their genes—in their brains by means of symbolic languages. The information can be passed along from one generation to the next, just as the genes do. However, language was only the beginning of the human information evolution.

Humans now have the capacity to store even more information, outside their genes and outside their brains, for the generations of the future. No other organisms have developed the building, machine-making, and symbol-using abilities to the degree humans have. However, as the world's population increases, there will be more people to organize more matter. The more energy and matter that must be used in the organizing effort, the more energy will be irretrievably lost. The law of entropy must take its toll.

Unlike most species, humanity does not need as many living members to protect the arrangement of matter in their genes. Humans can preserve the knowledge of entropy control in their brains and outside themselves, and they can use their knowledge to control more entropy. Humans can have guidelines for their conduct based on what is known about themselves as a part of natural phenomena.

A current review of the literature on moral education, moral development, and values clarification, has not revealed any references to the laws of nature. This is strange, since humans are a part of nature, and since they use energy, matter and information to move. Therefore, it seems purposeful to analyze the human system from an entropy control point of view, and indicate any implications for ethical instruction in public education.

NATURE AND EDUCATION

> He is the incidental biological by-product of water, soil, air, and sunlight.
>
> —Harlow Shapley
> The View from a Distant Star

Not every human system thinks he is a part of nature. Some young students, and, I am sure, many adults, are not aware that they belong to nature. When these people are asked to respond to "What is nature?" they always answer with some version of the following: "Nature is everything except humans and what they do."

If students are left ignorant of what they are made of and how they function, then we as educators are not doing a very thorough job. When students do not understand how they get their information from the world outside themselves, then is it any wonder they do not know they are a part of the environment? If our students do not understand the universe and where they are in the scheme of things, then how can they make valid decisions regarding their future actions and world problems?

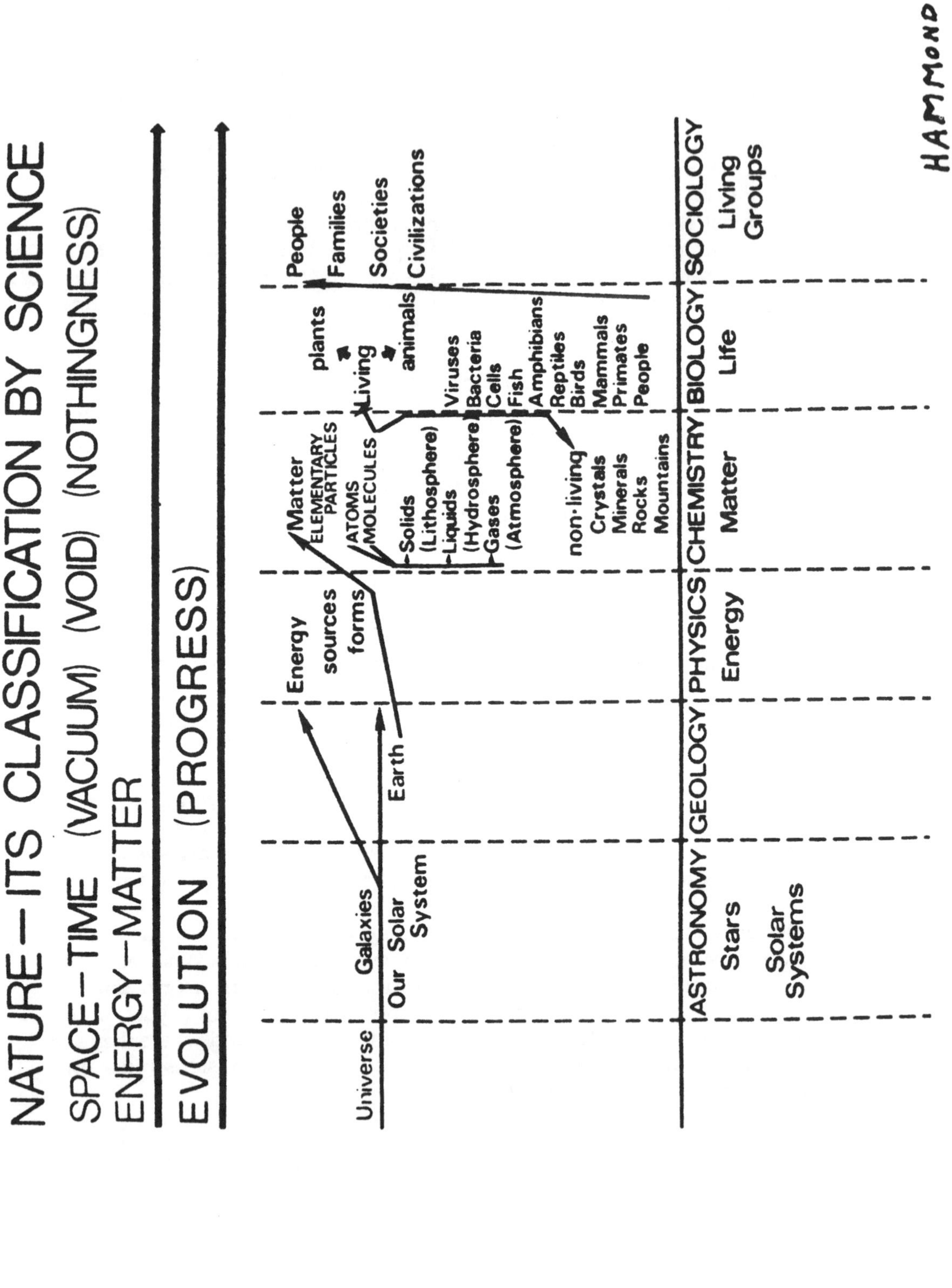
NATURE — ITS CLASSIFICATION BY SCIENCE
SPACE—TIME (VACUUM) (VOID) (NOTHINGNESS)
ENERGY—MATTER
EVOLUTION (PROGRESS)
Universe
Galaxies
Our Solar System
Earth
Energy
sources
forms
Matter
ELEMENTARY PARTICLES
ATOMS
MOLECULES
Solids (Lithosphere)
Liquids (Hydrosphere)
Gases (Atmosphere)
non-living
Crystals
Minerals
Rocks
Mountains
plants
Living
animals
Viruses
Bacteria
Cells
Fish
Amphibians
Reptiles
Birds
Mammals
Primates
People
People
Families
Societies
Civilizations
ASTRONOMY
Stars
Solar Systems
GEOLOGY
PHYSICS
Energy
CHEMISTRY
Matter
BIOLOGY
Life
SOCIOLOGY
Living Groups
HAMMOND

History tells us what happens to human systems when they think they are not a part of nature. They wreck, rape and ravage the rest of nature. The narrow view of thinking one is outside of nature makes the rest of nature something to be conquered, with total disregard for the consequences.

What every human system needs to know is that we humans cannot win any battle against nature. We can only be somewhat successful if we live in harmony with all of natural phenomena. Nature's laws will not yield. Thus, the human system must know what those laws are, to be fully aware of what to do as he goes through the process of living.

The human system is fixed in evolution the same as all other living systems. Also genetically, we reproduce our species the same as all other species do. What we probably have to a greater degree than any other species is an awareness, consciousness, or realization of where we are in evolution, and how reproduction works.

With this rise of awareness in the human system comes a responsibility to understand the fact that all information exists in the arrangement of matter. And even more importantly, the human system eventually understands he must deal with all the problems associated with reproduction. Who shall have sex with whom, how, when and where, and for what reasons? What about contagious venereal diseases? What about overpopulation? What about genetic engineering? If the human system does not think he is a part of nature, then no matter. The system simply takes care of himself and nature can do the same. That is the way of unaware organisms on the planet.

The human system must constantly take in energy and matter to grow and to go—to survive. This is another factor he holds in common with all other living systems. But, again, once the human system discovers more and more about his nutritional requirements, he begins to consider what he should and should not eat and breathe. What is the precise amount of each element and molecule to make the system grow and go as perfectly as possible for as long as possible? How much energy is needed when and where in each part of the human system? What is the best way to get the energy to where it is needed at the right time? What about ingesting certain bacteria, viruses and poisons? Which drugs are good for the human system and when? Which drugs are bad and when?

If the human system does not think he is a part of nature, then no matter. The system simply eats or breathes whatever is at hand. This system will either live or die by chance, or because of the caring of other human beings. Oh, for the haven of ignorance we once shared with our lower relatives on earth. Ignorance truly may be bliss, but once the human system knows about natural phenomena and what it takes for good breathing, good bones and good blood, then comes the responsibility to consider pollution, contamination in general, as well as the proper use of vitamins and minerals. Canning and storage considerations become implicated, as do all of agriculture, fertilizers, pesticides, waste disposal, littering, chemical dumping and water purification.

The human system has evolved five senses to monitor the outside world. These attributes he also holds in common with our closest relatives in nature.

These receptors account for information input into the human system's brain. They allow all experiences, and our educational endeavors must be built around these sensing mechanisms.

Again, once we learn the importance of seeing, hearing, touching, tasting and smelling, and how these processes work in our interrelationship with the rest of nature, then we realize that we are dependent on the rest of nature, more than the rest of nature is dependent on any single human system.

We know where we are in the universe because of the electromagnetic radiation energy coming to our planet from the rest of the cosmos. This is the second clue that information exists in the arrangement of matter. The first clue mentioned was the genetic information. The information existing in the arrangement of matter in the genes accounts for all living forms and their internal functions.

We also know all other objects in our environment according to the arrangement of matter—from the printed letters on this page, to the family photos, other films, forests and furniture.

Hearing is also dependent on how the molecules in the air are arranged. If there is no matter in the air to be organized, then there can be no sound. So this is an important third clue that information exists in the arrangement of matter.

The same is true for the senses of smell, taste and touch. Information is stored in the arrangement of matter, and the information the human system receives is thus dependent on the way matter is arranged—from genetic information to Braille for the blind.

This concept of matter arrangement and information storage may well prove to be as important a universal law of nature as gravity, velocity, density, or the direct relationship between energy and matter—$E = mc^2$

The human system may also conclude that it is highly probable he stores information in the brain in some manner which involves the arrangement of matter—however different from a computer, a book, or even a whole library.

Another natural phenomena must be taken into account when reasoning about the arrangement of matter and its relation to the human system. As matter combines and more complexity occurs, different characteristics appear. These different characteristics are called the states of matter. We shall review these states from the most jumbled and randomized, to the most ordered and organized.

The most randomized state is called a plasma. The inside of a star like our sun is hardly a fit place for the human system to inhabit. The rapidly and randomly moving elementary particles of matter in the sun would instantly scatter any organized matter sacrificed to it.

However, in a cooler environment, such as one here on earth, the elementary particles can stay together, so that atoms can remain more stable. Helium and the other noble elements remain very stable and are the least flexible. So it is the more flexible atoms which are needed to take us up the ladder to higher complexity, more diverse characteristics, and greater organization.

For example, hydrogen is our simplest, versatile atom. When it combines with another hydrogen atom, they form the lightest molecule, and the state of matter is a gas, H_2. But these same two atoms can also combine with one oxygen atom, thereby yielding a completely different state of matter, H_2O. Within given temperature boundaries, in which we live here on earth, water can be in the liquid state, whereas H_2 cannot. If we go up our ladder a step higher, we find that water molecules can combine with gaseous carbon dioxide molecules, CO_2, and a solid state occurs, sugar, $C_6H_{12}O_6$. The plants perform this task remarkably well, trapping the energy from the sun in the process we call photosynthesis.

As matter combines it acquires different characteristics, takes on different forms, and as I have noted, is the basic foundation from which we receive our information. We see these different characteristics, different states of matter, and different forms in the inanimate world around us. Not only do we gain information from the arrangement of matter beyond our planet, but also from our earth-bound atmosphere, hydrosphere and lithosphere. We see other living systems building inanimate structures which yield us information. And, of course, we see the human system reconstructing matter all over the surface of the planet, for better or for worse, yielding more and more new information.

In the inanimate world, matter reaches a high state of order in an object like a diamond. The human system seems to have a fondness for even inanimate objects which are in the highest state of order, far from the kind of randomness going on inside the sun. The stability of the crystalline structure allows a diamond to exist for a long period of time. This is the top of the ladder of organization for the inanimate world.

On the other hand, the ladder from randomized matter to highly organized matter for a single individual living system is far from stable, and therefore not very long lasting. But, oh how flexible, diversified, complex, interesting and informative these living systems are! I doubt if a diamond knows a thing about where it is in the universe. But the human system with its highly organized three pounds of matter, the brain, has reached a level on the ladder which is very high, indeed, and may go higher. By evolving through complexity and flexibility, the state of matter reached so far, by humans, goes rightfully by the name of awareness.

When we know these laws of nature, and how we interrelate with the rest of nature, we see clearly the human system as a part of nature. If the human system wants to educate, or to be educated then he must work with, and as a part of, natural phenomena.

The word unnatural does exist in the English language, but it is often used in a confusing manner. The term is implied or misused by advertising to sell products to people interested in 'natural' foods, cosmetics and the like. What does it mean to say; "This product contains no unnatural ingredients" or "This product contains only natural ingredients"? What is usually being avoided are the terms 'artificial,' 'synthetic,' or both. A way to fool the public?

Unfortunately many textbooks do the same thing—misuse the word nature and imply that humans, and what they do, are not a part of natural phenomena.

A hand-carved, marble elephant is not unnatural. It is an artificial elephant. Neither is a plastic elephant unnatural, rather it is simply an artificial, synthetic elephant. There are certainly differences between a beaver built dam and a human built dam, but we are misleading our students if we teach them one dam is a part of nature and the other is not.

Imagination and creativity may be marvels of the human system's brain, but they are by no measure unnatural. The gods of Mount Olympus or a heaven exactly three universes away from our own, are not unnatural. Neither are unicorns, mermaids, leprechauns or Santa's helpers. They are just as much a part of nature as the human system and its very wonderful, flexible and fertile brain, which gave rise to these possibilities. Here we simply need to teach the difference between fiction and nonfiction. They are attributes to be nurtured and cultivated, for without them we have no arts or sciences.

If the human system thinks he is a part of nature, then using the senses to the fullest will be considered very important. An awareness of input into the brain also leads to a realization of responsibilities. What more can be done for the blind? Who needs glasses and when? What about hearing aids? What can students learn with microscopes and telescopes?

If the human system does not think he is a part of nature, then he will certainly stay in, and perpetuate a state of ignorance. And ignorance is exactly the opposite of what education and science are all about.

The human system says and does things to himself and to the rest of nature. He must use energy from an outside source to behave. This he also has in common with our closest relatives in nature. Behavior is the process of using energy.

Again, once we understand the process of using our energy, we find that we become concerned with responsibilities. Why not use our energy any way we want to? If nothing is unnatural, why can't we burn down the forests, the art galleries and the libraries? Those behaviors are all a part of nature!

Where do we get guidelines for human conduct? How do we teach guidelines for human behavior? The human system, aware that he is a part of nature, becomes concerned for all of nature and, therefore, ethics. What are the right and wrong ways to behave? Or, stated better, what are the right and wrong ways to use our energy? The aware human system also wants to know why.

When the human system knows he is a part of nature, he carries the burden to help solve these complex problems concerning human behavior.

If a person thinks he is not a part of nature, that there is no kinship with nature—no interrelationship with nature—then that person disregards the rest of nature, and shows no responsibility toward the rest of nature. He will take care of himself without nature. He will say, "Nature is everything except humans and what they do." Such a human system will not be unnatural, but he will be wrong, and he will be ignorant.

Ignorance, for earlier humans, and even today for young humans, was bliss. Bliss, because nonawareness left them free from any responsibilities to the rest of nature.

The rise of science and formal education instilled an awareness in the human system, and presented a need for responsible behavior, which cannot now be ignored. The more science discovers about natural phenomena, and the more education teaches about the human situation in nature, the more burden is placed on the brain of the human system. There is a responsibility for the human brain to take more and more into account, when attempting to solve more and more complex problems. Is it any wonder that some students rebel against education in general, science courses, and science in particular?

For Further Reading

Cloud, Preston. 1978. *Cosmos, Earth, and Man*, Yale University Press.
Shapley, Harlow. 1963. *The View From A Distant Star*, Basic Books, Inc.

POSTSCRIPT

In nature there are five primary structuring processes which aid human existence. They can keep energy from scattering so rapidly, and they can keep matter arranged for a period of time, giving it form. This form which results from organized matter then provides information to the human senses. In this respect, the structuring of matter and information are the same entropy control process.

It must be made clear that any decrease in the orderly arrangement of matter is a step in the direction of more entropy and less information. The rapid and random motions of the elementary particles inside the stars cause them to have extremely high temperatures. And because of the chaotic motions and disorder within these plasmas, information about any particular particle is reduced to a minimum. However, these small pockets of plasma are by no means randomly distributed throughout the universe. The stars are contained in galaxies. Galaxies are shapes which can last a long time, being held together by *gravity*.

Gravity can bring about a long lasting structure, though undoubtedly not forever. So because of the orderly arrangements of the stars in the galaxies, and because energy moves away from the stars in accordance with the second law of thermodynamics (the entropy law), humans receive information about the galaxies. Indeed, the galaxies are known by their shapes, spins, measurements, motions, etc., not to mention the fact that *Homo sapiens*, at least, know where they are in the universe because of this information.

The same can be said for the solar system which is a highly stable arrangement of matter. Its very predictable motions, held together by gravity, not only have a store of obtainable information and thus entropy control, but the structure and function of the system has been the clockwork for all of life on earth.

Lastly, gravity can be thanked for keeping humans and all other unattached systems of matter from flying off the earth. Gravity helps keep entropy in line.

Moving from the most macroscopic consideration to the most microscopic, the nucleus of atoms are held together by *nuclear forces*. The elementary particles in the center of atoms are stable for some period of time here on earth, except for some minor and relatively orderly transformations. Because of the consistency of atomic nuclei, humans have been able to organize the periodic table of the elements, and the atoms of these elements have yielded much information. The nuclear forces do not allow the elementary particles to dissipate outward randomly, and thereby these forces help control entropy.

The third entropy control mechanism is the *electromagnetic attraction* which holds atoms and molecules together in chemical bonding, and accounts for the magnetic phenomena. Because the elements bond together in orderly arrangements, their attractive forces give rise to everything from a simple hydrogen gas molecule to the rigid structures of crystals, minerals, rocks and mountains; and to the more flexible, self-reproducing systems of all living things. The attractive force in chemical bonding controls entropy.

Plants are complex, reproducing, open systems which exchange energy and matter with their environment. Because they are highly organized and nonrandomly formed, *plants control entropy* by concentrating energy and organizing matter, thus they impart much information. Their importance in storing energy from the sun is well known, but unlike the higher animal life forms on earth, the plants have a minimum of individual mobility. They have little chance to organize more matter, or control more entropy, beyond the results of the genetic information passed on from one generation to the next in their chromosomes.

Like the plants, individual animals' bodies must grow in an entropy controlling manner or they will not survive. However, animals have the added ability to use their incoming energy not only to grow and to take form and to move about, but also to manipulate more matter and energy outside themselves.

Animal behavior is the fifth force capable of controlling entropy. And if an individual animal, clan, race, or species moves toward more and more randomizing behaviors, it will not survive. Be they paramecium, pelicans, or people, individuals move, by genetic and learned information, from less organizing to more organizing behaviors. People are entropy controllers and progress is a measure of the amount of entropy they control.

SCIENCE AND EDUCATION

> Some myths . . . explain how the world came into existence, the origin of mankind and the animals and plants, the origin of the two sexes, how human beings acquired fire, how society began, how work, old age, disease and the last enemy, death, came into the world.
>
> —Richard Cavendish
> Mythology

For scientists and we educators nature is the entire universe, including all of life on earth, the human system and what he does. As I have tried to make clear in the first chapter, it behooves all of us in education to accept this holistic view of the human system's involvement, and interdependence with the rest of nature. We cannot avoid this obligation or we will certainly mismanage the education of the students placed in our trust.

One of the roles of science is to find out more and more about the rest of the universe, our immediate solar environment and, indeed, the human system itself. Education emulates science in its own research, and is, therefore, one with science in any attempt to find out more about the educational process. However, both science and education have more practical roles to play than the pure research ones. Most scientists and educators spend more time teaching than ferreting out new knowledge.

Science, by its very activities, accumulates knowledge about natural phenomena and has provided humans with an enormous amount of knowledge. A very important role of education is to pass that knowledge on from one generation to the next, in the best way possible.

Science has also become a classifier of natural phenomena. (See chart #1 at the beginning of Chapter 1.) It should be noted that within the disciplines of science, all animate and inanimate forms and functions, have been put in order.

It must be remembered that at the very foundation of science is the human system. Each scientist, like each of our students, must be free to learn. He must be free to investigate the human system and all else in nature.

Through the human senses it is learned that relationships exist in nature. For example, velocity is a result of the relationship between time, distance and direction. Density is a result of the amount of matter in a given space. Information is a result of matter and its configurations. Entropy is a result of energy-matter and time. The faster the energy scatters in a given system, that is, becomes less concentrated and more dispersed, the faster it is rendered unavailable for further work. The faster matter scatters in a system, that is, becomes less arranged and more randomized, the faster it is rendered unavailable for further information. (See diagram #4 to find further relationships.)

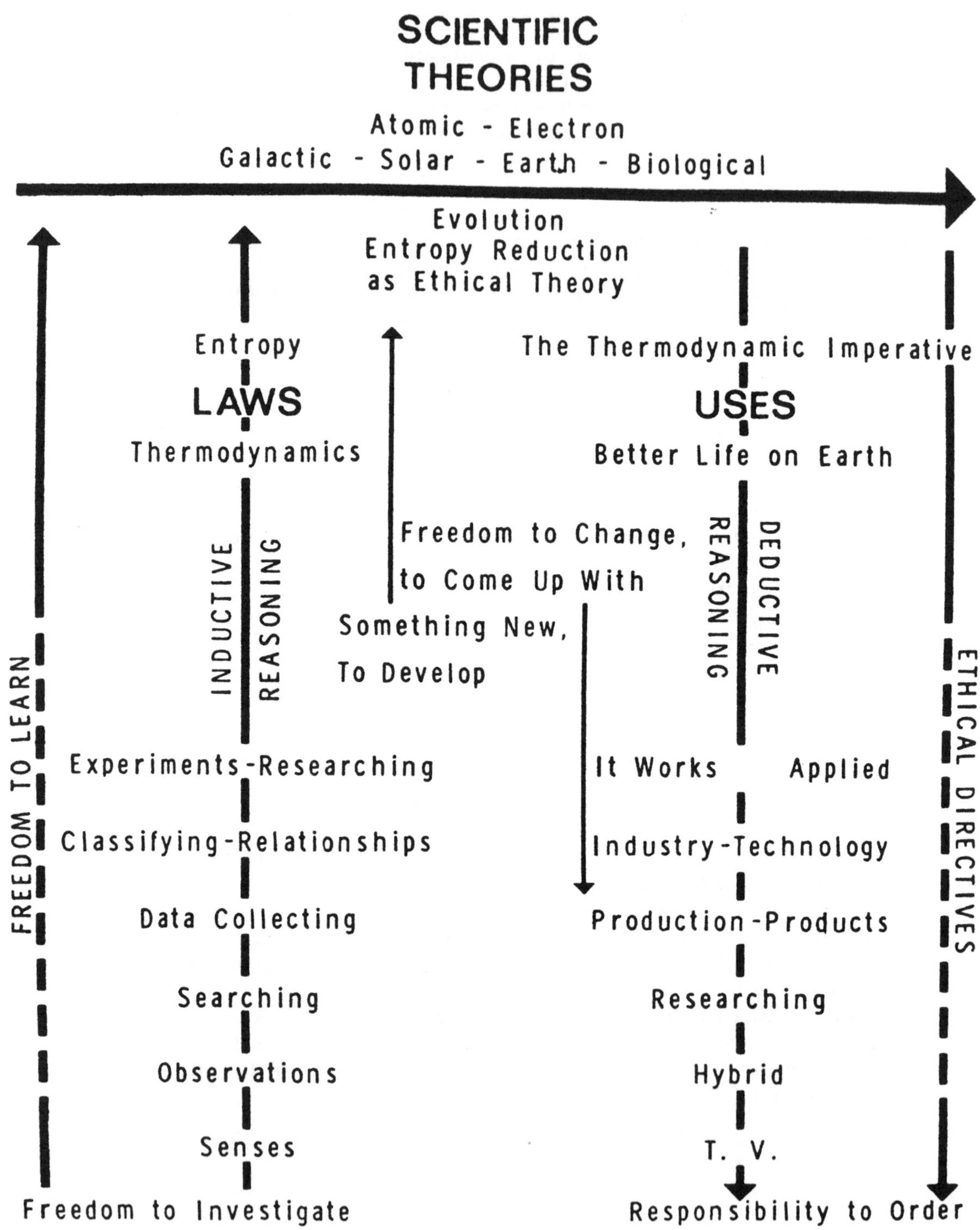
SCIENTIFIC THEORIES
Atomic - Electron
Galactic - Solar - Earth - Biological
Evolution
Entropy Reduction
as Ethical Theory
Entropy
LAWS
Thermodynamics
INDUCTIVE REASONING
Experiments-Researching
Classifying-Relationships
Data Collecting
Searching
Observations
Senses
FREEDOM TO LEARN
Freedom to Investigate
Freedom to Change,
to Come Up With
Something New,
To Develop
The Thermodynamic Imperative
USES
Better Life on Earth
DEDUCTIVE REASONING
It Works
Applied
Industry-Technology
Production-Products
Researching
Hybrid
T. V.
ETHICAL DIRECTIVES
Responsibility to Order

When scientists find relationships in nature which always hold true, then equations can be written. Numbers can be applied and problems solved. Such permanent relationships are granted the status of laws—the laws of nature discovered by scientists.

So far so good, because the units which make up a law can all be observed with the naked eye. But students and often teachers get into trouble in the next-higher-realm theories. Theories cannot be seen with the naked eye; rather, they must be derived by logical reasoning.

No one gets a concept of the Big Bang theory by staring at the stars in the sky. Nor can students understand or grasp the significance of the electron and atomic theories simply by observing a telephone. And, the theory of genetic biological evolution is far out of reach for anyone who only concentrates on a sequoia tree, or even tries to compare it to a giraffe.

The human system should go through the process of inductive reasoning, from using the senses, through the relationships and the laws of nature. And often, the relationships from several disciplines must be mentally put together to induce the theory, as is the case in the theory of genetic-biological evolution.

Teachers who cannot or will not guide their students through the necessary steps leading to an understanding of scientific theories, are leaving the students entrusted to them in a state of underdeveloped mentality.

One would think that in our highly scientific and technological society, scientific literacy would be one of education's primary goals. Not so, thus far. Students are graduated from high school and even college without a thorough understanding of what science has discovered about natural phenomena and their place in nature. This is bad enough, but there is more.

Education and educators failing to lead students up the mountain of knowledge to scientific theories leave students ill-prepared to distinguish between scientific theories and personal or group beliefs, or even myths. As with fiction, beliefs and myths should be clearly distinguished from scientific theories.

Along with inductive reasoning, scientific theories have a second pillar of support—deductive reasoning. If the theory is correct, then applied scientists should be able to use it.

Example: Nobody has been beyond our solar system to view its totality, or to photograph the entire system. But if the scientifically induced concept of our solar system is correct, then the technologists and engineers should be able to do something with the theory--like send spaceships to any object in the solar system and predict they will arrive where they are sent.

Again: Nobody sees or takes photographs of the total hydrologic cycle or the energy from the sun driving it. But if the scientifically induced concept is correct, then humans should be able to tap into the cycle, build dams and trap some of the water and energy for our own use.

And again: Though no one can photograph or directly observe what is going on inside the total electrical system of a car, if the atomic-electron theory is correct, then the lights, etc. should work quite well. In fact without the atomic-electron theory there would be no television, computers, calculators or as much knowledge of the human system and how it functions inside.

And again: The theory of genetic-biological evolution stretches over billions of years. Just as do the ages of the earth and the universe. But scientists did not need a motion picture camera to record all the events. The human brain became quite capable of putting these pieces of the puzzle of nature together. And again, if genetic-biological evolution is correct, then humans should not only know how and why living systems reproduce, but also how to control the breeding of plants and animals through time, including the human animal.

Thus, the two pillars of support for scientific theories are very strong indeed, and become even stronger as more knowledge is accumulated.

Education has the responsibility to pass along, from one generation to the next, this wealth of knowledge that science has accumulated about natural phenomena. And each student who is deprived of this knowledge becomes a mental cripple in our highly scientific and technological society.

Education also has the responsibility to make valid comparisons between the human systems' earlier mythological beliefs of earlier human explanations for natural phenomena. The myths and beliefs of earlier humans are rich in the history of human mental growth and development. They show that the human system has always needed an explanation for natural phenomena. One might assume that each human system still needs an explanation today.

But the day is past for any mythological or belief system to stand alone as a foundation for education. In our society it should not be the case that beliefs and myths are presented as being as valid as the findings of science. Remember we call scientifically illiterate nations underdeveloped. If that is true, then scientifically illiterate people in our own society must be classified as underdeveloped too.

In our society there are still groups who pressure the schools to equate their personal beliefs with scientific theories. And most pressure has been applied on the theory of genetic-biological evolution. If education were to succumb to such pressures, the we would weaken our nation scientifically and technologically. We would also not be presenting the difference between beliefs and scientific theories in a valid manner – a dishonesty. And worst of all, we would be setting up confusion in the minds of our students.

For Further Reading

Cavendish, Richard, editor. 1980. *Mythology an Illustrated Encyclopedia*, Orbis Publishing Limited.

Kneller, George F. 1978. *Science as a Human Endeavor*, Columbia University Press.

POSTSCRIPT

Education can only have a firm philosophy with science as one if its foundations and with science as its explanation for natural phenomena. Education should no longer continue to express its basics as merely reading, writing and arithmetic. Listening and speaking should certainly be added to that list, as well as understanding natural phenomena.

Many science organizations wanted to see education produce a decade of "scientifically literate' students in the 1970s. That desire was never fulfilled. The high school science courses were left as electives, and, as always, the mass of the students shied away from studying matter and energy (chemistry and physics). Only the best and the brightest, future engineers, doctors, dentists, architects, etc., elected to study and understand nature's basic laws and theories.

Of course not every student could handle the mathematics required for the junior and senior chemistry and physics courses. But were the courses on matter and energy for the non-mathematically oriented students–the majority of the students? Better to let those students go without this basic knowledge? Students without an adequate mathematical background had no alternative courses in science for them, no further science course requirements for them, and, I must add, not enough science teachers for them.

So, while other highly technological nations on earth required their high school students to take four years, a full load, of studies into the life, matter and energy domains, we in the United States left the bulk of our students ignorant, or at most only partially literate, in these areas.

Thus, as the 1980s began, our nation found itself with an acute shortage of science teachers, as well as basic research and applied scientists. The science teacher shortage stretched from pre-school through the doctorate to the highest university level. The many branches of government and all manner of private industries had tens of thousands of openings for all kinds of research scientists and engineers. At the same time, tens of thousands of people, ignorant about the basic laws of life, matter and energy, lined up for their welfare checks, because there was no work for them. Education has an obligation to become a buccinator for science.

INFORMATION AND EDUCATION

Whenever we put together a large, ordered structure from simple, randomly disposed units, whether it is a macromolecule or a brick wall, energy is required.

– Bioenergetics
Albert L. Lehninger

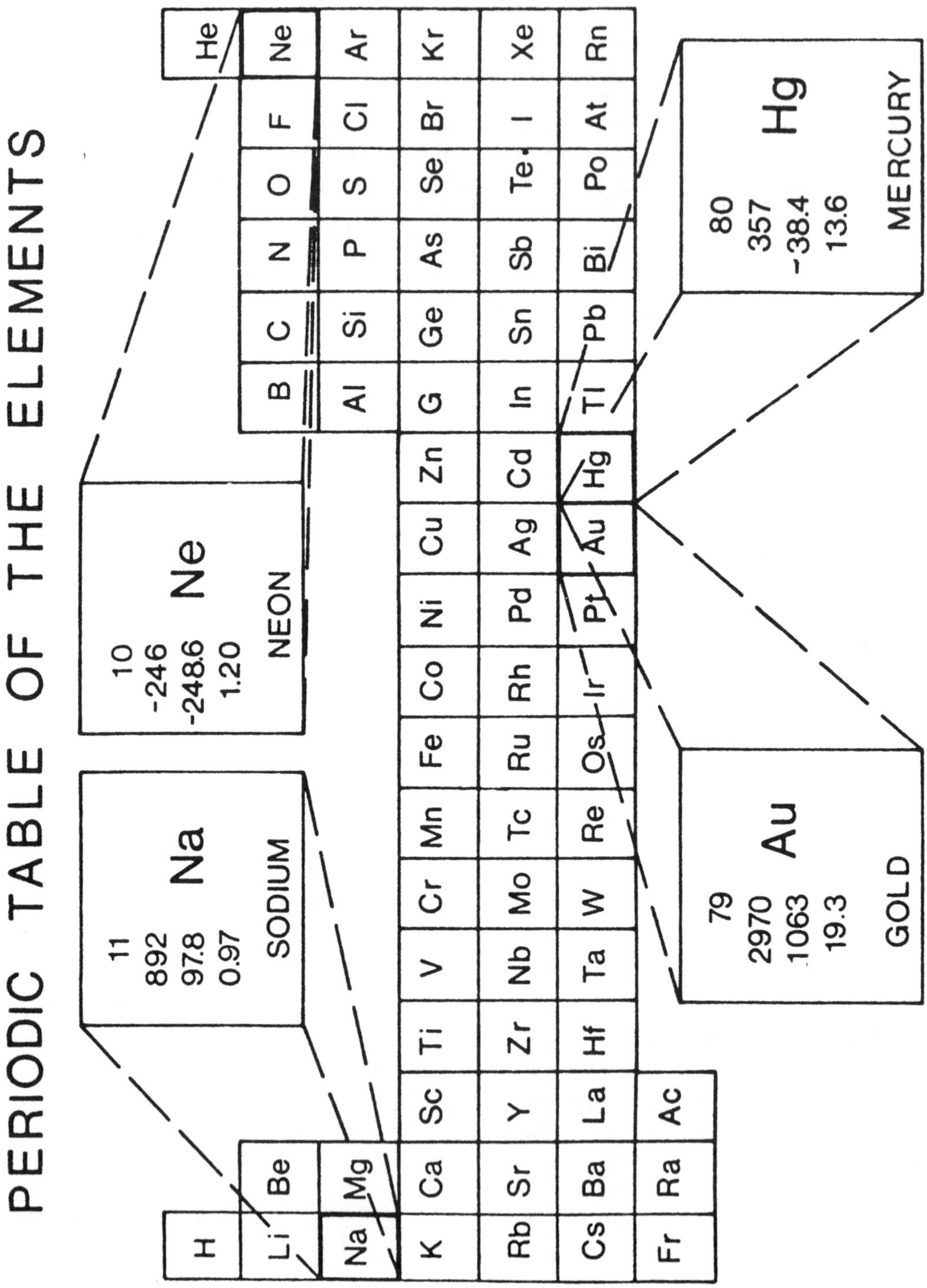
PERIODIC TABLE OF THE ELEMENTS
11
892
97.8
0.97
Na
SODIUM
10
-246
-248.6
1.20
Ne
NEON
79
2970
1063
19.3
Au
GOLD
80
357
-38.4
13.6
Hg
MERCURY
HAMMOND

As I have stated several times before, information exists in the arrangement of matter. Now it is necessary to clarify those terms, because the concept of "arrangement" includes more than one phenomena in nature, just as matter usually includes more than one element. This clarification is also necessary to explain why information theory should be one of the supports of a foundation for education.

Let us begin with examples from one of the most basic concepts in atomic theory. From the periodic table of the elements, scientists and science educators the world over teach their students that the atomic number is determined by the number of protons in the nucleus of the atoms. Each of these elements has its own properties and structural potentials.

For instance, if we look at sodium, element number eleven, it has eleven protons in its nucleus. At room temperature sodium is a metal. Though more than ten times less dense than lead, both sodium and lead can be cut easily with a knife. Of course, sodium has many other characteristics. When it comes in contact with water molecules an instantaneous and very volatile reaction occurs. Students are fascinated to see the sparks fly when a tiny piece of sodium is dropped into a beaker of water. This is a lesson and concept I find they remember, though the chemical formula for the reaction may somehow get lost in their recall mechanism.

Anyone wanting to store sodium in its pure state must take its violent reaction with water into account. Therefore, industry, research labs and schools must keep sodium sealed in a liquid other than water.

Imagine a small piece of sodium metal at room temperature. If a single proton from each of its atoms is removed, the decrease from eleven to ten protons transforms the soft, solid, metallic sodium, into a gas, and neon is formed. The newly transformed neon atoms will not react with any other atoms. (Though neon gas can be used by humans to cause energy transformations, yielding us neon light.)

It is interesting to note that the density of sodium is 0.97 less dense than water, which is 1.00. Thus sodium does not sink in water. It skips around on the surface of water, spitting sparks and causing waves. On the other hand neon has a density greater than water. Its 1.20 density allows it to sink below the surface of water. Yet it never reacts with the water molecules or any other atoms or molecules suspended in the water.

It should also be noted that the solid sodium metal will not melt to a liquid until it is heated to 97.8 degrees Celsius. And it will not turn to a gas until it reaches 892 degrees Celsius whereas neon does not liquify until its temperature is lowered to -246 degrees Celsius. When its temperature drops 2.6 more degrees Celsius, it solidifies.

I have compared sodium with neon to show that two atoms which have only the slightest structural differences can have enormous property differences. Though neon and sodium are neighbors on the periodic table, their characteristics and the information they yield are vastly different.

There is another example I want to use from basic atomic structure. Gold and mercury are also neighbors on the periodic table. They too are very close to having the same structure in the arrangement of their elementary particles.

Gold is element number seventy-nine. It is pretty, solid, yellowish, shiny and metallic at room temperature. Gold is useful and humans have placed great value on it. If one more proton is added to each of the atoms in a piece of gold, it is transformed into mercury, element number eighty.

Mercury is a liquid at room temperature, existing in the form of balls which can easily be made smaller or larger. They all look alike except for their size. Students are fascinated with mercury because of its unique properties. But, of course, it should be learned (hopefully not by trial and error or personal discovery) that mercury too must be handled with care. It can be deadly to the human system.

We may also note that mercury's temperature must be lowered to -38.4° Celsius before it will solidify, while gold will not melt until it reaches 1063° Celsius. Mercury will boil if its temperature is raised to a mere 357° Celsius, whereas gold must be raised to 2970° Celsius before it boils.

These examples, at the most basic level of atomic structure, show that matter displays itself to us in a wide diversity of characteristics, which yield the human system much information. Continuing up the ladder of matter, from elementary particles and atoms to molecules and macromolecules, we will see why structures, properties, characteristics, the arrangement of matter, and thus information become limitless.

As atoms join together, greater complexity and more flexibility become possible. The molecules exhibit a wide diversity of states and shapes. The information matter yields for human assimilation becomes infinite. This is true in the realm of organic molecules (those carbon based), and in the realm of living systems.

There are two basic phenomena involved in the macromolecular build-up to higher levels of complexity. The first necessity is chemical bonding, or put more simply, atoms attaching to other atoms. If that phenomena did not occur, there would be no molecules, large or small; no crystals, minerals, rocks or mountains; no water, no life, etc.

The second phenomena is exemplified by the isomers. They are molecules which have the same chemical formulas, the same number and kinds of atoms; however, they are arranged differently. An example is to compare ethyl alcohol with dimethyl ether. Their molecular formulas are both C_2H_6O. But the carbon, hydrogen and oxygen atoms are arranged differently, causing the molecules to have completely different properties.

Ethyl alcohol is a liquid at room temperature, and dimethyl ether is a gas. Dimethyl ether does not react at all with sodium, but ethyl alcohol reacts vigorously with sodium, yielding hydrogen gas. Ethyl alcohol will not boil until it reaches 78.5° Celsius, but dimethyl ether boils at -24° Celsius. Still each has the same number and kinds of atoms. Thus our only conclusion can be that since the matter is arranged differently it has different properties, and therefore yields us different information.

To show why isomerism helps account for infinite matter arrangement and thus infinite information, let us look at one more simple, though one step higher, formula from organic chemistry. $C_{40}H_{82}$ can exist as a simple carbon and hydrogen chain of atoms. But there are also known to be 6.25 x 10^{13} total arrangements possible for this molecular formula. That is 62,500,000,000,000, or 62 trillion 500 billion. Still not infinite, but there are several million carbon compounds known to exist, and carbon is just one element on the periodic table, though admittedly one of the most flexible. If we add one more flexible atom, silicon, and all of its possible compounds, then we are far beyond any possible counting by the human mind.

The possibilities for information from matter arrangement are uncountable, and as incomprehensible as are any infinite amounts. This is true for the animate as well as the inanimate realms of nature.

Natural disasters threaten the organization of matter and information. Natural disasters rapidly dissipate energy to an unusable state, and even threaten the survival of living systems.

The basic organizers of matter are gravity, the nuclear forces, chemical bonding, and living systems—so few to construct and create so much. For when we observe the deteriorating and destructive forces in nature, we see the challenge that evolving life on earth has had to contend with to stay alive.

The human system, like all other animals, plants, and organized systems on earth, has had to cope with the destructive forces of wind and rain and snow and floods. People have had to repair their structures after the effects of hurricanes, typhoons, tornadoes, dust devils, tsunamis, tidal waves, earthquakes, avalanches, erosion, freezing and thawing, volcanoes, lightning, and fires; not to mention the debilitating work of lice, rats, mice, termites, worms, mosquitoes, fungi, bacteria, viruses, poisons, rusts, and other sources of decay and deterioration.

Destroying organized systems will always be an easy task. However, creating new and better organized systems is necessary for the survival of life on earth. It would seem that the ethical instruction of the young must therefore be to teach them that their role in life is to create, and to help maintain entropy reducing systems.

Is it any wonder that the striving to protect and pass along the information in the genes is the foundation for evolution—for bacteria or bushes, for beetles or bushmen? Is it any wonder the human system has tended to worship the creator and punish the destroyer?

In spite of the destructive and debilitating forces omnipresent in nature, the plants climbed the ladder from simple chemical bonding to complex macromolecules by trapping the energy from the sun. Thus, with energy from an outside source, they have been able to produce the carbohydrates, proteins, lipids, starches, oils, terpenes, waxes, steroids, alkaloids, pigments, cellulose fibers, gums, pectins, resins, approximately 15 billion tons of wood, and about 25 billion tons of carbon dioxide each year. Is it any wonder that the aware human system appreciates the processes, products, properties, and information the plants pass along to us?

The plants trap the sun's energy, contain it in chemical bonds, and thus concentrate it, so the energy ceases to scatter and becomes usable. Thus it is that the plants on earth are able to locally reduce entropy. Entropy dictates that energy flows to a more scattered state, and matter becomes less organized. Plants use energy to construct matter from more randomized states to highly organized states. Thus, they are entropy reducers in this sense also.

Finally, by arranging matter in an orderly fashion, the plants are capable of creating an infinite variety of information. Entropy is related to information inversely. The more information, the less entropy. The plants are entropy reducers in an information sense, also. Any wonder that the aware human system hates the ignorant human system who irresponsibly burns down the forest? It seems imperative that education should instill an entropy reduction ethic into the minds of oncoming generations.

As we continue with the higher organization of organic matter, we see characteristics and states of matter and processes which can only exist by more complexity. Sameness and diversity of each individual living system, when compared to all other living systems, is a good point to consider.

Within each species the coded information in the genes causes each member of the species to be and look different from every other species. (There are some rare exceptions due to the evolution of camouflage for survival.) Also, within each species, like our own, each one of us is and looks different from every other human system. (One exception is one-egg identical twins, and another might be extremely rare look-alikes.) Yet, we all seem to be very similar when we compare ourselves to grasshoppers, giraffes or grass. So, how does this diversity within sameness come about?

To answer this we must look at the large molecules of which the genes are a part, and specifically at the information points along the DNA molecule that are called codons. They contain the precisely arranged matter (instructions) which are replicated when cells and new chromosomes are formed for cell division. In this template-like, self-replicating process, what are the chances for the codons to come up with a human system who will look different from each one of us?

The human system goes meticulously and precisely through this process in sperm and egg production. So we should look to the sperm and egg cells, where the chromosomes are halved, to find out what the possibilities are for each new human system to be unique.

Each human system, male or female, can produce 2^{23} genetically different sex cells. That is 8,388,608 different sets of information for each past, present and future male or female. Since it takes the information from one egg and one sperm to start a new human system, any man and woman can start 70,368,744,177,664 different kinds of human systems. Multiply that times one-half of all the humans who have lived and will ever live and we see that human variability is infinite. All possibilities for life on our planet are truly incomprehensible.

Any wonder that speculating people think there is a high possibility that an infinite number of life forms could exist in other parts of the universe?

The human system has evolved the five senses and the brain to experience and assimilate an infinite variety of information. Science and the many other human endeavors which have classified and organized natural phenomena, expect education to pass this knowledge on to each new human system in the most organized and efficient (easily learned) way possible.

If we fail, the new generations will be ignorant and confused about their place in nature. They also may not learn what is acceptable behavior, as they go through the process of living. Even more important, if the educator fails, then the students may not learn a firm foundation for ethics, which tells each student why some behaviors are good and some are bad.

A final point must be made about information. It is the arrangement of matter which controls the flow of energy. If energy is to be stored rather than scattered; if energy flow is to be slowed down or speeded up; or if energy is to have its direction of flow channelled, then the organization of matter by humans is imperative.

The plants discovered this simple fact long before any humans were around to become aware of it. Oh, that the human system could trap the scattering energy from the sun and store it so easily! However, like the other animals, we cannot eat direct solar energy. For our own bodily growing and going, we are completely dependent on the entropy decrease taking place inside the plants.

Like the work of the chlorophyll in the plants, the information in the genes of all living systems build the systems (arrange the matter) in order to control energy storage and flow. There is no other way to existence for life. Fortunately, we do not have to think about building our own circulatory systems, or any other of our subsystems which make us whole. What a task that would be!

However, as the human system becomes more energy-wise, we are seeing the necessity for education to place more and more knowledge about energy into the curriculum. We now realize that the information entering the brain (via energy transmission) has enormous control over how we use our energy in behaving. In fact, any considerations about human behavior are considerations about how the human system uses its energy. So to ask any ethical questions like "How should the human system behave?" is the same as asking, "How should the human system use its energy?"

With energy as a valuable factor within the human system, and in its behavior, an energy control foundation for ethics is emerging. That is why the subtitle for this book is *From Entropy to Ethics*. The topic of guidelines for human behavior, from what science has discovered about natural phenomena, will be treated in depth at the end of this book.

For now, it should be made clear that humans have learned to control energy, organize matter, and thereby yield the human system new information by emulating what the genes do—by constructing more and more complex systems. In human evolution, information storage has expanded from the genes, to include the brains, and now to the systems that humans build.

For Further Reading

Holum, John R. 1979. *Elements of General and Biological Chemistry,* John Wiley and Sons, Inc.

Lehninger, Albert L. 1973. *Bioenergetics,* W.A. Benjamin, Inc.

POSTSCRIPT

In this essay I have dealt with the most basic aspects of information—the arrangement of matter yields us information and determines our form and function, as it does for any other living system.

What makes humans so different from other life forms is their ability to arrange matter outside themselves, and to recognize it as representing something which it is not. Pictures, films, records, photos, models, graphs, charts, words, numbers, etc. are an infinite variety of symbols to represent the rest of nature. A painting of three trees is not three trees. The words "three trees" are not three trees. And the number "3 trees" is not three trees. Rather, they are all ways for us to communicate with each other about trees. They are ways of symbolizing and classifying nature, so that nature becomes manageable to the human brain.

Since there are an infinite number of forms and functions, properties and relationships in nature, we assemble into groups and categories those representations which will be the most beneficial to each oncoming generation. We start with each child's immediate self and environment. This is what each society, each culture, and each family does for their new generations by means of their language. It is necessary that within each language, nature becomes classified and manageable for communication and ultimately for the education of the human system.

Education is responsible for human beings acquiring higher and more complex communication skills. But the infinite number of symbols cannot simply be dumped randomly into the brains of children. Education also has the task of organizing what goes into the brain for the better building of communication skills. This is the difference between instilling confusion and understanding. We educators must help each student understand the symbols which stand for other natural phenomena. This is how the tree of knowledge, awareness, and logical reasoning grows best inside the human brain, and it is because the roots of knowledge are in the arrangement of matter, the residence of information.

Another point should be made about information. The arrangement of matter can be destroyed.

When a library is burned down, or when a tapestry is eaten up by moths, or when a child's quilt is dissolved in acid, the energy and matter of which the items are made are not destroyed. There is no way to create or destroy energy or matter. They can only be transformed, one into the other, and moved about from place to place in the universe. But the buildings, the books, the symbols, the pictures, and the numbers, in other words, the information, can be destroyed.

The organized matter is what makes the world around us manageable and meaningful. When this organized matter and concentrated energy, (information and locally reduced entropy) are lost from our further usage, we do not weep because the energy and matter have been dissipated and dispersed. Of the billions of people who have existed, most of them have not understood that energy-matter cannot be created or destroyed. But still they have had to make organized matter and organizing matter very valuable to them. It should remain so, and is one of the most basic values educators pass along to their students. Why?

1. Because of communication, as we have seen.
2. Because the cost of replacing organized matter can be very great in terms of more money, time, knowledge, matter and energy required.
3. Because the immediate loss of organized matter can threaten the very existence of the loser.

Deterioration, decay, destruction and death all remind us how fragile existence is. Better to punish the destroyer and worship the creator!

SYSTEMS AND EDUCATION

Wholes are not the simple sum of their parts, and heaps are.

—Ervin Laszlo
The Systems View of the World

A pile of rubbish is an arrangement of matter, but it cannot be classified as a system. A randomized mound of glass, paper, metal, plastic, dirt, rubber, wood and whatever else, has no holistic, overall organized structure. Something more is needed for matter to become a system.... a whole.

If energy as heat, light or sound were to be passed through a heap of debris, the energy's flow would be scattered and unpredictably randomized. If more matter, such as a rag, or a used bottle cap, or a piece of hard clay were to be added to the junk pile, the additional matter would not change the unorganized nature of the trash. They would simply add to its size and mass.

Only energy moves in and out of a closed system. A music box is a closed system. It is organized matter which controls the energy transformations, storage, rate of flow, and direction of flow.

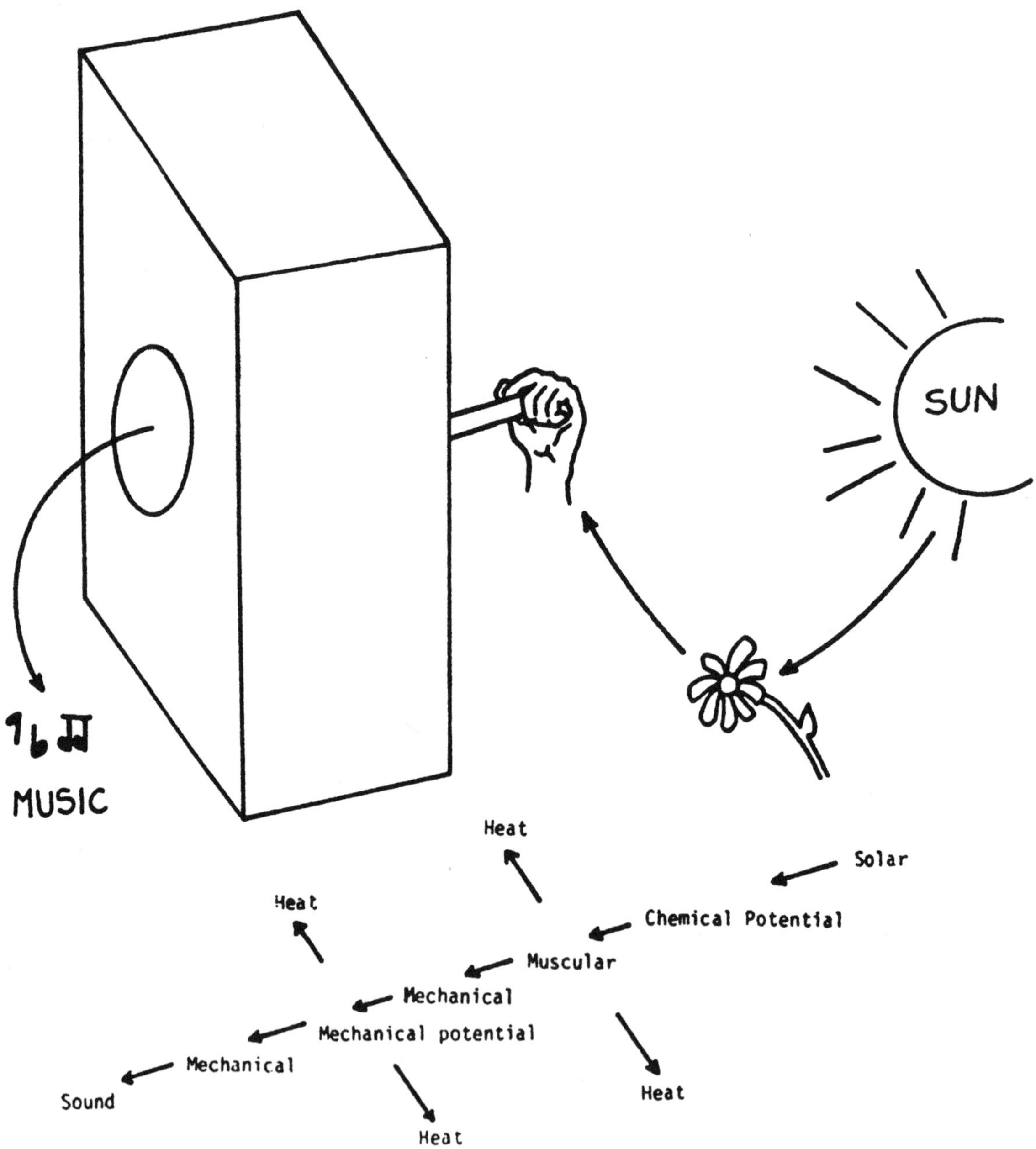

Organized matter is more than the sum of its parts, heaps or not. When matter is organized energy flow is predictable and the flow can be controlled, as diagram #5 shows. This is one of the most important differences between a system and unorganized matter–energy control.

By energy, I mean the stuff that comes to us from the sun. I am concerned with the same energy that astronomers and physicists think about, write about and calculate–the stuff that all stars emit because of the nuclear fusion going on within them. This is the same energy that geologists and chemists must take into account to understand how our planet's atmosphere functions or how photosynthesis works. Without a knowledge of energy the biologists couldn't explain how animals are able to move about, nor could the technologists have built the world's elaborate transportation systems. Without energy, nothing grows and nothing goes.

The physical and biological scientists have come to realize that humans, like all other animals, are energy dependent and are just as much a part of nature. Since energy is so necessary for the very existence of plants and animals on this planet, perhaps the human behaviorists should say more about energy. Because we are having an energy crisis on earth, perhaps it is time for the psychologists, the sociologists, the educators, the religious, and other people concerned with ethics and human behavior to include more information about energy in their thinking and in their evaluations. The physical and biological scientists also have need to include more reasoning about human behavior in their energy considerations.

We know where we are in the universe because of the energy arriving here from the stars in our own galaxy, as well as from the galaxies outside our own. We are able to follow the flow of energy from the sun to our ability to move. We can follow the energy used in playing the piano to the ear of the receiver, and on into the brain. We diagram energy's route from the sun, through photosynthesis to fossil fuels, and on to the electric generating plant, and from there into the home where it toasts the bread. We have knowledge about how to get energy from one place to another. We have information about energy transformations and energy laws.

Should only the physical scientists, the biologists and the engineers be concerned with energy? Or do we all need to know that energy from the sun and the other stars is the same stuff, the same entity that comes from the center of the earth and the center of atoms? Should we all know how energy flows, where it comes from, and where it goes?

Isn't it necessary for each person to realize that energy and matter are the same thing, like two sides of the same coin? Like the sides of a coin, we see energy and matter as two separated substances, but we should know they are transformable, one into the other. That is why scientists talk of energy-matter as though they are one–the substance of everything we see in the universe. The rest (if, indeed, we can give it a name) seems to be an infinite void, a nothingness we call empty space. Because we recognize the flow of energy-matter in space, we are able to construct a measurable concept of time, to which we attach a one-directional flow. Thus it is that the universe seems to be a finite amount of energy-matter which exists and flows from one place and form to another in an infinite space-time.

This brings us to the first law concerning energy and matter. ENERGY-MATTER CANNOT BE CREATED OR DESTROYED. Of all the experiments done by humans, no one has ever been able to detect energy-matter appearing from nothingness. And, no one has been able to annihilate energy-matter, that is, send it to nothingness. So until someone finds an exception, the first law holds; energy-matter can neither be created nor destroyed.

The first law is taught and kept in the minds of physical and biological science students from elementary school through the most advanced graduate college courses. More often than not, the first law raises more questions for the thoughtful student than it answers. Since we educators are in the business of developing questioning minds, the questing of the students should be seen as sparking their curiosity. In many students the first law can even cause consternation.

Students ask and exclaim:

> Where did all this energy-matter in the universe come from? It must have come from something! Even the law itself says that something does not come from nothing! But if something did create the energy-matter in the universe, then that something must have been created by something! Oh, that line of reasoning can go on and on forever!

Philosophers throughout the ages have called this kind of thinking infinite regress.

The students continue:

> Well then, are big bangs going to go on and on forever? Or are the galaxies going to move away from each other forever? What is this mess of thinking we're into? Must all of us who contemplate the first law of energy-matter live without final answers? Must science classes always take us up the mountain of knowledge, only to reach the top, and leave us unknowing again, with no higher point to climb? Is the job of science, then, to build the mountain of knowledge higher?

"Yes," I finally answer my students, "that is the task science has set for itself, to accumulate more knowledge."

It is exciting to watch young minds grappling with ideas, and maintaining their curiosity. They have reached the pinnacle of understanding science and how it works. When students realize how much humans *do not* know, how much work there is to be done to increase the mountain of knowledge, then perhaps they will opt to help science accumulate more knowledge.

In the future it will be necessary for all students to understand the first law. Like money and time, information and matter-energy are accountables, not to be left out of human reasoning.

Often the first law is stated rather whimsically, "There is no free lunch." This simply means that humans do not know how to create energy-matter from nothingness. Students soon realize that if somebody could disprove the first law, a Nobel Prize would be in the offing. There would be a "free lunch" for all of life on earth.

But for now, humans must pay a higher and higher energy price, not only for their lunch, but for the more noble and socially acceptable, and therefore ethical, ways of behaving. As more humans go about the business of accumulating more knowledge, which means utilizing more organized matter, more energy will be needed. Education needs systems for efficiency to occur in learning.

When we think in terms of systems, like the metric system, we think of simplifying and structuring for better organization and understanding. We try to bring about more order and more efficiency when we think of some entity or some process being more systematic. We categorize all of natural phenomena, from the objects and their functions to all of human behavior, and even what goes on inside the human body and brain.

We look at the world around us in terms of systems. It is a common word in our vocabulary. Witness: "I see our spiral Milky Way galaxy as a system of billions of stars"; "Today we build a model of our solar system," "There is a low pressure system developing to the southeast of us," "The mass transit system in our town doesn't work well," "I just had the electrical system in my car fixed"; "He says there's something wrong with his digestive system"; "There must be a better registration system"; "This tree has a deeper root system"; "We have a new computer system in our office"; "Let's think about the electron system in this molecule." These examples simply remind us that we look at, and talk about, all manner of things in nature, in terms of systems.

From the smallest hydrogen atom, to the entire universe, we communicate about natural phenomena in terms of systems. We recognize that there are non-living systems and there are living systems. Just as energy is identified by its forms, matter is categorized on the periodic table of elements, and life is divided into the plant and animal kingdoms; so too systems must be put into groups for easier understanding.

We also need to integrate energy, matter and information as accountables into our systems considerations. This is not difficult because there are only three basic kinds of systems. They are *isolated*, *closed* and *open systems*.

An isolated system would have no energy or matter passing in or out of the system. Within our rapidly moving and energy dispersing universe, perhaps no pure, isolated system really exists. But the human mind is capable of imagining isolated systems. Like absolute zero, infinity, and traveling at the speed of light, people can contemplate such situations, so they do.

Suppose we could place a perfect diamond in the center of a perfectly vacuumed container. Then suppose we could place this isolated system inside some perfect shield to keep out all energy and matter. Finally, the diamond in the vacuum, in the shield, is suspended at a point in space where no gravity budges it one way or the other. Now we've imagined an almost perfect isolated system.

Our imagined isolated system is almost at absolute zero, but not quite. The movements of the elementary particles inside the diamond cannot be stopped. There are also motions inside the atoms of the vacuumed container and the shield. Now we are close enough to look at some important features of an isolated system.

The closer a system gets to a perfect isolated state, the closer it gets to absolute zero, and the slower change will take place within that system. The hypothetical, perfect diamond will remain unchanged for a very long time. Since we only know time by change, for an isolated system at absolute zero, time stands still. The system exists at maximum stability. It has an unchanging order. In pure isolated systems no atoms can move about the container. If this happened, then some of their energy of motion would eventually be transmitted outside the system. It would then no longer be qualified for the isolated systems category.

If I could see inside the isolated system I have imagined I would see no motion. The energy-matter of the system would remain highly ordered; I would see the carbon atoms as "diamond." In other words, because the energy and matter of this system are not scattered randomly throughout the system, I am able to receive information about the system. The carbon atoms would yield the information, "diamond," for a very long time.

But could I actually see inside an isolated system? The answer is no. Seeing means light energy going to an object and then to my eye. This would mean energy going into and out of my imagined system. Thus it is that pure isolated systems can be imagined, but probably do not exist. If such a system did exist, we could not know it.

Isolated systems fall into the category with absolute zero chambers and perpetual motion machines. Here on this earth and perhaps in the universe it is probably very close to impossible for a pure isolated system to exist with no energy or matter passing into or out of the system.

What about considering the universe as a system? Again, it must be imagined, of course. Is there any matter or energy coming into or going out of the universe? If the universe is expelling matter or energy, where is it going? If the universe is receiving matter or energy, where is it coming from?

If the universe is an isolated system, then its energy-matter content is a constant. If the universe is boundless, then the void we call empty space is without limit. The empty, nothingness part of the universe seems infinite, having a bit of energy-matter within it which can neither be created nor destroyed, and constantly moves from one form and place to another.

As energy and matter change forms and flow, an inevitable phenomenon occurs. Left to itself, energy-matter becomes less concentrated, less ordered, more dispersed and more scattered, almost always tending toward maximum randomness and equilibrium. Without any energy, matter and information being applied to a system from outside the system, that system will go toward maximum disorder. This, of course, includes that largest of all systems, the universe, and all systems inside the universe, including living systems and the systems humans build. Thus we have the second energy law.

The rate at which a system goes to disorder is called its entropy. Thus it is that when a system's energy and matter are in a high state of randomness (the pile of trash), entropy is high. And when a system's energy-matter are in a high state of order, such as my imagined, isolated diamond, then entropy is very low. As we have seen, low entropy, or high order is also related to information. From a human point of view—and I might add, for all living organisms—as entropy increases information is lost and so is energy control. And conversely, as entropy decreases information can be gained and energy controlled. I will have more to say about this phenomenon in the following discussion of closed systems, and later in relation to education and ethics.

A closed system exchanges energy with its environment, but not matter. Here are two examples of closed systems which operate in nature independent of living systems and what they build. The tides gain their height, then ebb, by using gravitational energy. The hydrological cycle is essentially driven by two sources of energy—the sun and gravity. Within living systems, many familiar examples of closed systems are available at the submicroscopic, energy transmission level.

A nervous system takes in energy from outside itself, transmits that energy, and finally gives it up at some energy transformation point. A nervous system is a closed system, rapidly delivering information inside living systems via energy transmission.

If we think more broadly about communication, and include what goes on at the environmental interface with living systems, we find that the receptors for seeing, hearing, touching, smelling and tasting are simply transforming energy and not taking in matter. Living systems have evolved extremely sophisticated closed systems as a means of monitoring their internal structures and functions, as well as for monitoring structures and functions outside themselves.

Human technology has developed and improved the same kinds of closed systems for external communications that living systems evolved internally. Examples are the telephone system using wires, and the wireless use of the waves in the electromagnetic spectrum. Humans have learned to build these closed systems which will transmit energy from one place to another via solids, liquids and gases, or across the empty void of nothingness. Once they are built, energy input is all that is needed. (Electrons are considered as energy here, and their slight mass is disregarded for the sake of macroscopic discussion.) It is no longer necessary to introduce matter for them to function. Thus, they are closed systems.

Closed communication systems are able to function, according to the second energy law, because energy flows from an area of greater concentration to one of less. Without any help from outside a system, this one directional flow is always true, and, thus it is a law. All electrical systems operate in this same manner, and once built by humans, they are closed systems. They transmit energy from an area of greater concentration to one of less. What makes electrical systems different is that their purposes are often broader than communications. They are used for all manner of work at the ends of their transmission lines. An electrical system may carry energy to perform such frivolous tasks as driving staples through two sheets of paper, sharpening pencils, or burning toast. But the system can also be so important as to supply energy for the work of a pump, lighting the darkness, or monitoring a heartbeat during surgery.

Humans have learned to control the rate and direction of energy flow outside themselves. Closed energy systems, which exist outside the living systems, are indeed a mark of progress. They are technology at its finest hour.

There are many other examples of human-made closed systems which use the energy forms of heat, light and sound control. There are many examples involving the simple machines. Playground equipment, wind-up time pieces, toys and music boxes are only a few more illustrations of devices made to control the rate and direction of energy flow. They are closed systems which can be made to last longer than the living system which built them.

Unlike imagined, isolated systems, closed systems are real, and an infinite number can be built. In the realm of possibilities, humans have only built a few closed systems. Thus it is that engineering will remain a very vital and lucrative occupation.

When we look at the similarities between the above isolated and closed systems, we see that both human-built systems have their entropy reduced. With information, humans have learned to organize matter in such a way that the finished system regulates the rate and direction of energy flow. As we will see, perhaps the primary task of not only the engineers, but all of life on earth, including humans, is to reduce entropy. To exist, life must slow down the rate at which energy and matter disperse and scatter, tending toward maximum randomness and information loss.

It is the arrangement of matter which yields information, thus leading to energy control and entropy reduction. As we now move into an analysis of open systems, the growth and development of living systems, and especially humans, we will begin to see the far reaching implications that this total concept has for the purpose of education and the foundation of ethics.

For Further Reading

Laszlo, Ervin. 1972. *The Systems View of the World,* George Braziller.

Emery, F.E. (editor). 1974. *Systems Thinking,* Penguin Books, Ltd.

POSTSCRIPT

Musical instruments are closed systems. Initially humans arrange the matter in building the instrument. The human muscular energy is used to create organized sound waves among the atmospheric molecules. The percussions, strings, brasses and woodwinds are all systems humans have devised to control energy.

The most sophisticated closed systems built by humans are computers, robots, and their combinations which use microcircuitry. Complex android systems have receptors which perform the same functions as the human senses. Their energy for function is controlled by the way humans build them (arrange the matter in putting them together).

An open system takes in and expels both energy and matter. Living systems are the most sophisticated open systems on the planet earth. Open systems, built by termites, birds or humans are, as yet, very crude when compared to what genetic evolution has been able to build.

Open systems evolved because chemical potential energy proved, by photosynthesis, to be an excellent way to trap, hold, and transport energy. Then at the site where the energy is needed (whether in the human muscles or the automobile engines) combustion releases the energy. But, as we will see, there is a very fundamental difference between living systems and combustion engine machines as open systems.

A combustion engine vehicle can only use the energy it takes in. But since that energy is bound up in chemical bonding, the matter intake is only incidental to the energy flow and overall functioning. Thus the matter taken in must be gotten rid of as soon as possible after the energy is released. The matter taken in would rapidly ruin the system if allowed to stay inside. Thus all combustion engines are open systems only because of the necessity to eliminate (by the exhaust system) the unwanted matter.

So the open systems built by humans, or any other animal on earth, cannot yet use the matter taken in to grow or to repair and maintain themselves. (Oh, that our vehicles could mend themselves the way our bodies do!) That is the basic difference between the two kinds of open systems, the non-living and the living.

THE HUMAN SYSTEM

Every living organism is essentially an open system. It maintains itself in a continuous inflow and outflow, a building up and breaking down of components, never being, so long as it is alive, in a state of chemical and thermodynamic equilibrium but maintained in a so-called steady state which is distinct from the latter. This is the very essence

> of that fundamental phenomenon of life which is called metabolism, the chemical processes within living cells.
>
> —Ludwig von Bertalanffy
> General System Theory

The human system is an open system. From conception to death the human system must take in, use and expel matter and energy to be able to grow and to go. As in all living systems, it is the genetic information which controls the energy and matter utilization, leading to a highly organized system. Since this is so, the human system, like any other living single cell, plant or animal, can be studied from a system's point of view. The human system can be looked at where its four primary domains interface with the rest of natural phenomena. These four domains are:

1. The genetic-evolution domain.
2. The energy-matter input domain.
3. The sensory-education domain.
4. The energy-matter output domain.

As we have seen, if there is not some structure to matter, some organizational arrangement in matter, then there can be no energy control. In other words, without the organization of matter (which is the same as information) the energy laws tell us that energy rapidly disperses to an unusable state. Energy scatters from an area of high concentration to one of low concentration. We see this one directional flow from a burning candle, from a sound source, from any mechanical device, from an electrical system and, fortunately for us, from the sun.

If that tendency to flow toward equilibrium is to be reversed, then additional energy and matter must be used from outside the system. In addition, know-how, which comes from information and reasoning, must be applied.

Human engineering performs this feat quite well, as can be observed in the example of the refrigerator-freezer. Just plug in the electric cord of the refrigerator and the heat energy inside is moved to the outside. Inside, the entropy will decrease. A tray of cold or frozen water molecules has more order, less entropy, than the same hot water with molecules rapidly evaporating. But outside, all the energy leaves as heat, and some unwanted sound, going out into the space around the earth. The energy is lost to further human usage, except for warming the kitchen a bit along the way.

The point is that humans have learned how to reduce entropy locally. The know-how comes from the human brain. We see that even a simple closed system like a heat pump can have its internal entropy reduced. But, as we are all aware, the refrigerator, like any other functioning device built by humans, will eventually wear out and deteriorate.

THE HUMAN SYSTEM

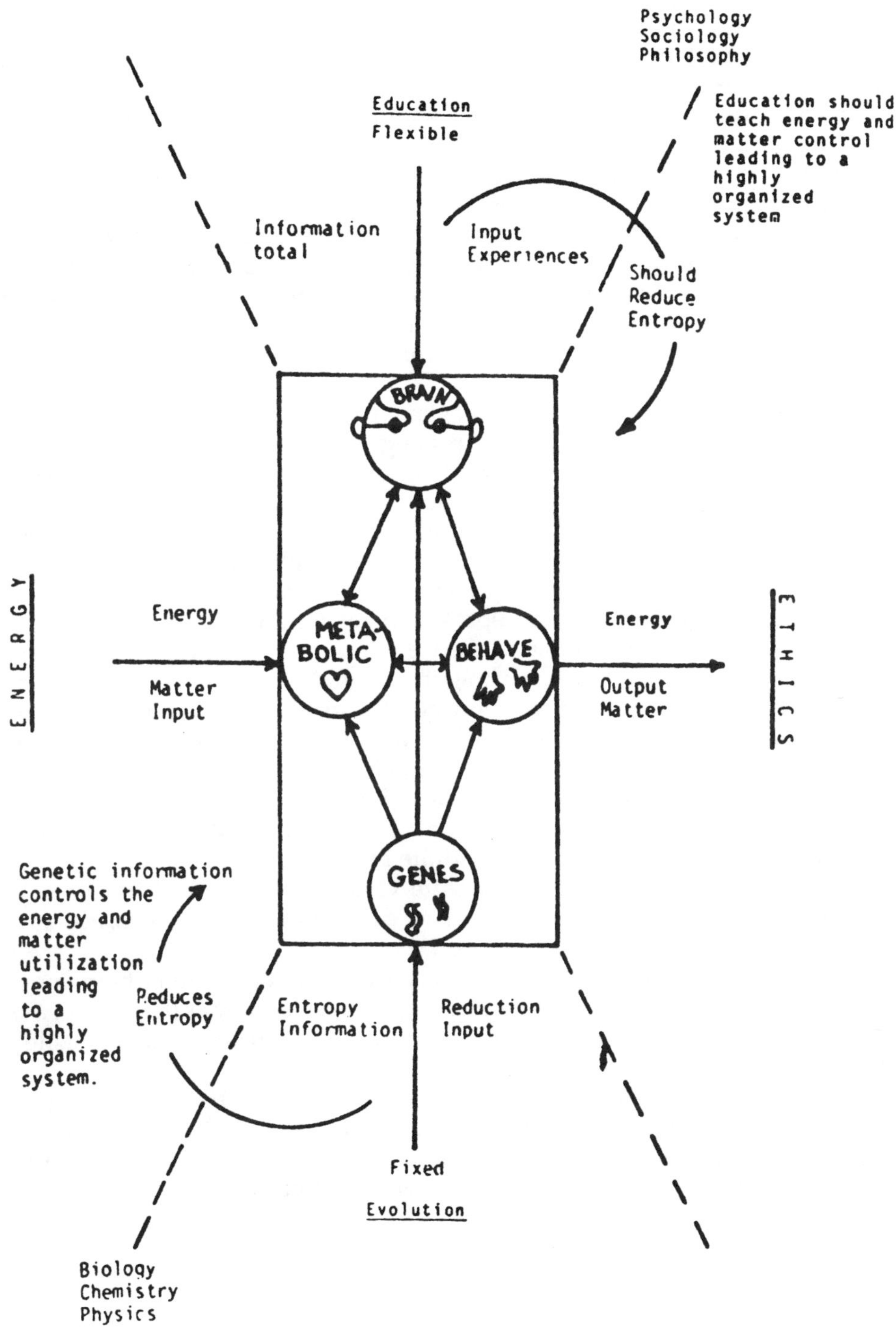

This is precisely what all living systems have had to contend with. They have had to control the flow of energy to exist, because they have had to reduce entropy to exist. But unlike the education needed to develop the engineer, the genes evolved the same know-how billions of years ago, with an added dimension not yet simulated by human engineering. Since any individual functioning system will eventually wear out, something more is needed for prolonged survival. The genetic solution was, and still is, to assimilate usable elements into living systems, along with energy. This process has led to an infinite variety of open systems. Therefore, unlike the closed systems built by engineers, the living, open systems are able to make internal repairs.

But there is still another aspect of the living, open system which is just as important as internal maintenance, and perhaps even more so. The information in the genes must be reproduced for each cell in a living organism, and it must also be passed along from one generation to the next, else the whole process of reducing entropy would cease. We see in genetic possibilities constantly changing evolution built into each one of us, into each species, and indeed into all of life on earth.

Because the organizational arrangement of matter (information) is the key to prolonged survival, it is not surprising that the brain has evolved to such prominence in the human system. The brain evolved enormous flexibility, unlike the fixed genetic information each one of us inherits.

Throughout nature there are infinite possibilities for the arrangement of matter. If the human brain were to cope with all the possibilities, the best survival tactic would be to have as few restrictions as possible on what information the brain could accept. When restrictions on sensory input exist, organisms are very vulnerable to their environment. One does not see or read about blind and deaf animals existing in the wild. Humans do take care of such offspring, but that is the exception not the rule. Plants and animals which cannot fully monitor their surroundings are subject to both the knowledge and desire of the human system to keep them alive.

We see the flexibility of the brain when we conceptually survey all the past and present human cultures. Think of the number of different languages. Because the brain is so flexible, an infinite number of written and spoken words are possible. They represent the infinite variety of forms and functions in the rest of nature. This accounts for the many different types of beliefs, customs and traditions expressed in human cultures.

Although each human system is very similar to every other human system, and in basic aspects to all living systems, we have seen that the arrangement of matter in the genes also allows for each one of us to be distinguishable from every other living system.

The brain is also evolving to allow each one of us to be different from every other human system, both mentally and behaviorally. The human brain, like the genes, is geared for adaptation and change ... a tested way to survival, though, perhaps a very difficult phenomena for educators to contend with.

The genes have been wending their way through the branches of biological evolution for billions of years. The genes found a way to utilize energy and matter for prolonged survival. Now the human brain is becoming aware that it too must work with the same natural phenomena, if the human system is to be successful in continuing to inhabit this planet.

For Further Reading

von Bertalanffy, Ludwig. 1969. *General System Theory*, George Braziller.
Darlington, C.D. 1978. *The Little Universe of Man*, Geor, Allen, & Unwin.

POSTSCRIPT

The human system uses energy to move. In other words, human behavior and energy utilization are the same phenomenon. One of the tasks of education is to get the human system to use its energy in a particular way. Educators ask for specific behaviors.

This striving to control entropy, to reduce or decrease entropy, is a necessity for all living systems on earth, and probably any other place in the universe. No human system is an exception. The value of order over chaos is built into all aspects of human biological growth, mental and behavioral development. The information in the genes causes the energy and matter input to fashion a highly organizing human system. The highly organized information input into the brain causes the highly organizing behaviors to be valued. Behaviors causing chaos are not sanctioned. Reducing entropy is the underpinning of all survival.

Governments, religions, industries, science, law, health, communications and educational enterprises operate within and outside their systems so as to move societies away from chaos toward greater order. The same is true for the arts. Neither a nation nor a social organization can survive for long in a state of chaos, turmoil, commotion and confusion.

As education prepares oncoming generations to take up their roles as acceptable citizens, there is a heavy obligation to reduce the chaos and confusion, both in the brain and in the behavior of each maturing human system.

Because survival dictates that entropy be reduced, managed and controlled, no nation or educational organization can shirk this basic responsibility. What is needed now is a better understanding of the total entropy paradigm and how it applies to the human system from entropy to ethics.

PART 3

The Entropy Ethic

INTRODUCTION

Possibly every social entity is constantly facing some new problem. This happens because we do not live in rigid stagnation, but rather, in constant change. How boring life would be without change. So we must take the bitter with the sweet and face the problems change brings with it.

One of the problems thrust on education in our changing, pluralistic, free, democratic society is the question of ethical education for the young. For more than ten years the public has been asking education for better ethical instruction in the public schools. There have been some responses to this demand, but little of any substance.

I have tried to understand why this is so. My research in the library has given me little help. I am left with piecing the puzzle together as best I can. So I will say, "it seems to me," and let it go at that.

In history "it seems to me" that education and ethics came out of the same womb—religion. Religion and ethics were slowly removed from public education following the scientific revolutions. Historians tie the first scientific revolution to Copernicus. That is when religions began to lose their power in explaining natural phenomena. Some have struggled to perpetuate their beliefs about nature for centuries, but scientific evidence has mostly prevailed over religious explanations.

In this domain of education about natural phenomena, the most devastating blows came to religions after 1850. In astronomy Copernicus was proven correct. Humans and their planet are not the center of the universe. Not only that, but we also do not go around a stationary star in neat little circles. And later with knowledge of the speed of light came the understanding of our expanding universe, and the fact that we cannot see any two things simultaneously in the cosmos.

After 1850 physics too was to gain enormous knowledge about natural phenomena. Scientists figured out the sources and forms of energy, the transformations energy goes through, and the energy laws. The entropy concept evolved from equilibrium thermodynamics to nonequilibrium thermodynamics of open systems, including humans and what they do.

Chemistry finally shook off alchemy and the old concept of matter being air, fire, water and earth, and their combinations. The periodic table of the elements was born. Matter would henceforth be molecules, their atoms, and the atoms' elementary particles interacting with energy.

In biology came more and more knowledge about how plants' and animals' inner systems operate. With an understanding of energy flow and photosynthesis came a sound explanation of how plants and the animals that eat them are able to grow and go. Vitalism was exorcised from the human system. In addition, the theories of biological evolution and genetics gained world-wide scientific acceptance.

Psychology strove to become a science and to shed itself of religious explanations. Souls and other non-scientific terms were replaced with terms in tune with scientific experimentation.

And finally, in sociology, the rise of communism would cause incalculable problems for religions. And, of course, religions would come to realize even democracy could exist without them, but they stood a good chance of not existing without democracy or their own dictatorships.

I admit the above are only highlights of happenings since 1850. There has been much more in all the sciences. But with merely that much said, "it seems to me" a very broad scientific revolution began in the 1850's. It is much bigger than the revolution Copernicus initiated, because this one involves all the science disciplines, physical, biological and social. And the revolution is still causing enormous consternation in the thinking of humans all over the planet.

A few fundamentalist religions continue their ways of explaining natural phenomena. And, of course, they are free to do so in our democratic society. But, "it seems to me" they could maintain more credibility by being objective about the scientific evidence.

What most religions have done is give up the domain of trying to explain natural phenomena, leaving that to science. Besides, religions had a much more important role in the history of social development, and they still do. Religions help to keep order in societies by offering a foundation for ethics. They tell people how they ought to feel about each other, and offer some right and wrong behaviors in the realm of interpersonal relationships.

When we expelled religions from the public schools in the United States, out went instruction in ethics too, creating the void we now have.

The above scenario may be oversimplified, but "it seems to me" it is essentially on the right track. What we are left with is a lack of formal ethical education, and the public is aware of this problem. What the entropy ethic can do for education is help fill that void.

After years of working with adolescents in and out of the classroom setting, there is another point I must consider here. Teenagers in the United States are very sophisticated in their thinking today. Mass communications, the home and the schools may have done a better job than they sometimes think they have. Adolescents have been taught to be creative, to come up with something new, to live with changes, to have curious, inquiring minds, and to question everything.

In our free, democratic, pluralistic society, a public school value, of necessity, has *not* been to blindly follow any one supernatural, dogmatic system of belief. Thus, we cannot expect the majority of the high school students to accept anything like the old "Bribe of Heaven and Fear of Hell" reason for behaving one way rather than another. That foundation for ethics is shattered.

My search for an ethical foundation in current psychology and human growth and development books reveals little that will satisfy the public, the parents, the teachers, or the teenagers. Other than a little dabbling into trying to find out what society and students think is valuable, textbooks for educators and students are shallow and superfluous when it comes to satisfying the adolescent mind in the area of ethics.

Then what are we to do as the public cries for better ethical education in the public schools? One place to begin is with the entropy ethic. Of course, that is what this part of the book is supposed to be about. So it is best that I get on with it.

In the fall of 1977 Ilya Prigogine won the Nobel Prize in chemistry for his work in non-equilibrium, irreversible thermodynamics of open systems. *Time* magazine called him "...a poet of thermodynamics..."

Many scientists and writers soon came to realize that Prigogine's major contribution was that he bridged the many chasms between the physical, biological and social sciences. As Mary Lukas put it in *Quest/80*, "Prigogine has taken a major step toward healing the ancient breach between the life and nonlife sciences, with implications that seem to grant humankind more control over its fate than many scientists ever thought possible."

Rifken and Howard immediately saw the potential of the entropy concept. They rapidly pounded to press *Entropy: A New World View*. The research was very thorough, so that their book contains one of the best bibliographies on the subject of entropy. The book predicts that, "Already the outline of the new entropy paradigm is being filled in by scholars around the world. Within a few years every academic discipline will be turned inside out by the new entropy conception."

On an ethical note, they also state, "The Entropy Law answers the central question that every culture throughout history has had to grapple with: How should human beings behave in the world?"

The entropy concept is a very powerful analytical tool for any discipline from astronomy and the arts, to biology and economics, to music and math, and on to physics and philosophy. And many scientists also see a new foundation for ethics inherent in the total entropy concept.

To get to the ethical components we must analyze what it takes to have a local, low entropy, living system, including the most flexible and complex of all living systems, *The Human System*.

ENERGY AND ETHICS

> So, somehow living organisms including plants succeed in reducing their entropy.
>
> —Richard L. Gregory
> Mind in Science

Looking at life on earth holistically, living systems use the sun as their main source of energy. Humans also use a little energy from other stars, as information for knowledge. And science has also learned to tap the center of atoms and the heat from the center of the earth to get a little stored energy. But again, for the most part our energy for growing, thinking, and behaving comes from the sun.

Like all other animals, humans have evolved to live off the plants, either directly or indirectly.

From an energy point of view, entropy tells us that energy flows from an area of greater concentration to one of less, scattering toward equilibrium, until it becomes useless. The trick to existence was, and still is, to trap the energy scattering away from the sun, and to concentrate it, making it usable. This is what all living things must do for survival. This is one sense of the fact that living systems are antientropic, consume entropy, or live off of negative entropy. By concentrating some energy and using it to bring matter to a higher state of organization living systems locally decrease or reduce entropy.

No work would get done inside a living system, and there would be no energy for behavior if energy flow was not controlled, and held out of balance in a nonequilibrium state. From an energy point of view it is a part of every living system to ward off entropic doom. There is no other way to existence. Efficient use of energy is an integral value for the body, and the brain, and the behavior of a living system. Looked at as a whole, the human system has a built-in ethical principle—*use energy efficiently*.

The Human System is an island of reduced entropy in a sea of confusion, corrosion, contamination, corruption, conflict and chaos. Our low entropy status reveals an ethical principle to us: We should use our personal energy as efficiently as possible to bring about order in our environment.

Using energy efficiently to build a new human system is not something we can do much about, so we need not worry about it. All other living systems seem to do quite well without even knowing about energy. Many social insects and other animals have this principle built into their genes which controls their behavior. Humans do too, but it has been masked by the evolution of the brain and the need for education.

The new human system is born without the ability to organize its environment. The psychomotor development of a human is a very long process. I will contend that in this process there is the underlying principle of the entropy ethic—the need to learn to utilize energy efficiently to bring about order in the environment.

The newly born human system is a little randomizer—uncoordinated. The efficient use of energy in psychomotor development is a long learning process. Randomizing, destructive behaviors are what each new human system begins with, and *NOT* what parents and schools hope to instill in each new generation.

Many animals, such as some insects, spiders and birds, have some fairly complex, built-in, genetic know-how for constructive, self-disciplined, psychomotor skills. The human system does not. Thus, our need for learning, conditioning for coordinated skills, and education in this all-important area of psychomotor development.

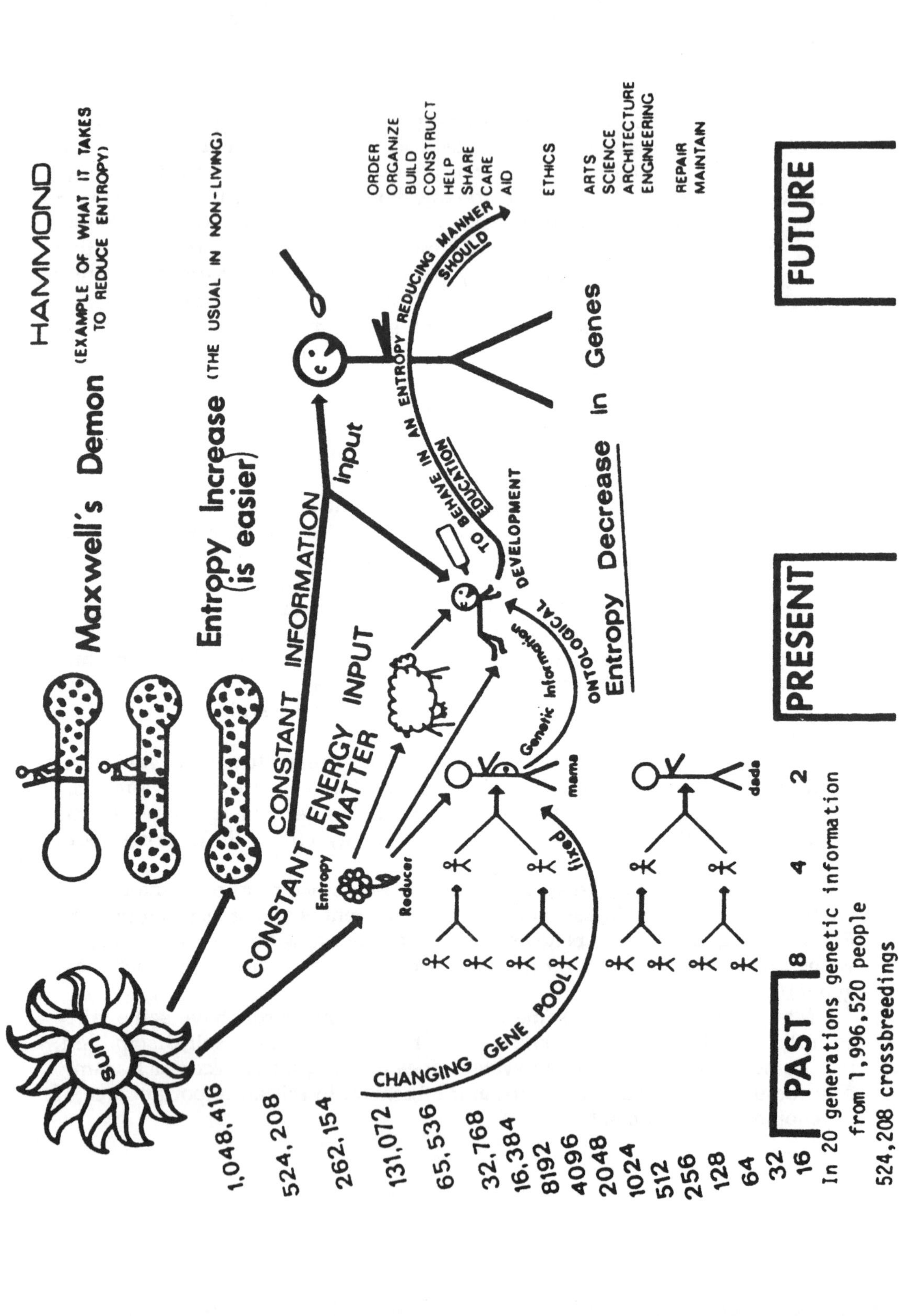
HAMMOND
Maxwell's Demon (EXAMPLE OF WHAT IT TAKES TO REDUCE ENTROPY)
Entropy Increase (is easier) (THE USUAL IN NON-LIVING)
sun
CONSTANT ENERGY MATTER INPUT
CONSTANT INFORMATION
Input
Entropy Reducer
TO BEHAVE IN AN ENTROPY REDUCING MANNER
SHOULD
EDUCATION
DEVELOPMENT
ONTOLOGICAL
Genetic information
(fixed)
mama
dada
CHANGING GENE POOL
Entropy Decrease In Genes
ORDER
ORGANIZE
BUILD
CONSTRUCT
HELP
SHARE
CARE
AID
ETHICS
ARTS
SCIENCE
ARCHITECTURE
ENGINEERING
REPAIR
MAINTAIN
1,048,416
524,208
262,154
131,072
65,536
32,768
16,384
8192
4096
2048
1024
512
256
128
64
32
16
8
4
2
PAST
PRESENT
FUTURE
In 20 generations genetic information
from 1,996,520 people
524,208 crossbreedings

One of the parents' and teachers' primary ethical goals is to guide the human system from behavioral disorder to behavioral order. If this is not accomplished, we say the student lacks self-control or self-discipline. And disruptive and destructive, disorderly and disorganized adult humans are not acceptable. The entropy ethic is thus built into the educational endeavor to help maintain social order.

To establish the necessary discipline, parents, teachers, counselors and administrators try behavioral modification, paddling, economic and freedom rewards, treating the student with dignity, and even sometimes positive reinforcement. No matter the method, the goal is always the same. We must help each human system acquire self-discipline. The value is order in the home and in the classroom, just as it is inside the human system and inside the human society at large.

The basic ethic of the medical professions is not different from that of the classroom teacher. They both strive to bring about and to maintain the order in the systems they work with. This is a worldwide basic ethic and another primary component of the entropy ethic.

Of the hundreds of recommendations I have filled out for future teachers, students and colleagues, they all place value on the organizer, the person with self-discipline, and the ability to bring about order in the environment. Employers do not want randomizers, who cannot use their energy efficiently, and who worship confusion, conflict, and chaos.

For Further Reading

Gregory, Richard L. 1981. *Mind in Science*, Cambridge University Press.
Rifkin, Jeremy. 1980. *Entropy: A New World View*, The Viking Press.

POSTSCRIPT

How should a person behave?

How should an open, human system use energy?

These are both identical questions. But the first gives no clue as to which foundation for ethics will be followed to arrive at an answer, no clue as to what are to be considered as accountables.

The second question is the same ethical question. However, it points directly to energy as a consideration, a path to follow to help find a solution. Please note that I could have, and will, ask the second question with at least three other endings in place of energy. How should an open, human system use matter, information and time?

All of these four, energy, matter, information and time are inextricably related to entropy, and I will treat each of them in that context in the following chapters.

For now, however, I want to add a few thoughts about energy, entropy, education and ethics, to show how they are necessarily related.

We educators do *NOT* allow the human systems in our trust to use their energy any way they want. We do, in fact, follow the innate ethical imperative. And it is anti-entropic. What we have *NOT* learned to do is to spell it out as simply, clearly, easily and concisely as, say, the medical doctor has learned to do. Thus, the parents, the future teachers, the general public, and particularly the students have a hard time understanding the value of what's happening to them in education.

I am going to use the example of teaching music (which I don't know much about, but many of my students do) to oncoming generations. My students have taught me that what I am going to say can be applied to any discipline in formal education, and even to discipline itself. Thus is the power of the entropy concept in analyzing education.

I begin with a question.

Some day why don't we just give a random assortment of musical instruments to any hundred teenage, high school students and let them put on a large orchestration for the public that same night?

Obviously, if we did that we would have a hodgepodge of unorganized, random sound waves—noise. And noise is very high entropy. It irritates and is unacceptable to human reason. True the human system's hearing mechanism will accept random sounds. Hearing and the other senses are very flexible in what they accept. It is the brain which rejects randomness and chaotic input. If confusion is what is put into the brain, then that is what comes out. Or as they say in computer slang, "garbage in, garbage out." Not much worth or survival value in that.

The base line is that noise or high entropy is not valuable to humans in any communication mode. Our hypothetical school has not put together an orchestra, but rather a mess of highly disordered sound waves. A rapid, local move to high entropy is not something we humans favor. Whether it is an untrained orchestra or typist, an accidental burning of a forest or an intentional burning of a library, destruction on the highway or any sudden premature death of a healthy human system, rising chaos is not a condition to which we attach high value.

Acceptable behavior is *not* killing other healthy human systems, *not* randomly destroying orderly objects humans build, *not* burning down libraries and forests, *not* typing randomly, and *not* creating chaos with sound waves.

Thus it can be seen that formal education is one of the social institutions which strives to dispel localized disorder and helps create social order. Education knowingly or unknowingly follows the entropy ethic. But it would be better if each educator understood what he is doing and why. Because only then will he be able to teach the entropy ethic.

Before going on to the next chapter, I must make a final comment on how many educators view a growing human system.

Educators have been trained to look primarily at the behavior of a student from kindergarten through high school. That seemed to make sense because behavior is easily observable and testable.

We can easily categorize the students by their age and what they can or cannot do. But these divisions we use of ages, stages, feelings, psychomotor, cognitive, and many more can be detrimental as well as useful. Looked at holistically, each human must grow and develop from a single fertilized egg to an open, low-entropy, complex, flexible, highly organized system. Speaking, walking, typing, and all other behaviors in general should not be viewed as detached or separate from the student. For in essence they are not. All parts make up the whole system. Any part going from order to disorder affects the whole human system. To live, the human system's parts, from the genes to the organs to the behaviors, must grow and go from disorder to order. That is the foundation for the entropy ethic.

The educator who sees reading or writing or the staging of a play or studying history, as some separated behavior of the human system is missing the mark of what the growth of a human system is *all* about, does not know the essence of what he is trying to do, and unfortunately cannot get to the entropy ethic.

MATTER AND ETHICS

> All men should fight always as vigorously as possible to increase the degree of order in their environment, i.e., consume as much entropy as possible, in order to combat the natural tendency for entropy to increase and for order in the universe to be transformed into disorder, in accordance with the second law of thermodynamics.
>
> —Robert Bruce Lindsay
> The Role of Science in Civilization

In the previous chapter I tried to dwell on each human system's personal energy control. How that energy flow-through is used for orderly body building and bodily behavior is paramount to education and to ethics.

The newborn human system cannot speak, write, stack the blocks, put the puzzle together, fix the car, build a hotel, or help another human system stay in a state of high organization. A baby is even dependent on other people for proper energy-matter and information input so he can grow and go in a low entropy manner.

It is *not* ethical to feed a growing infant material which will cause disorder to his system. The entropy ethic tells us that that would be against the very nature of individual evolution. So are beating, battering and bruising human

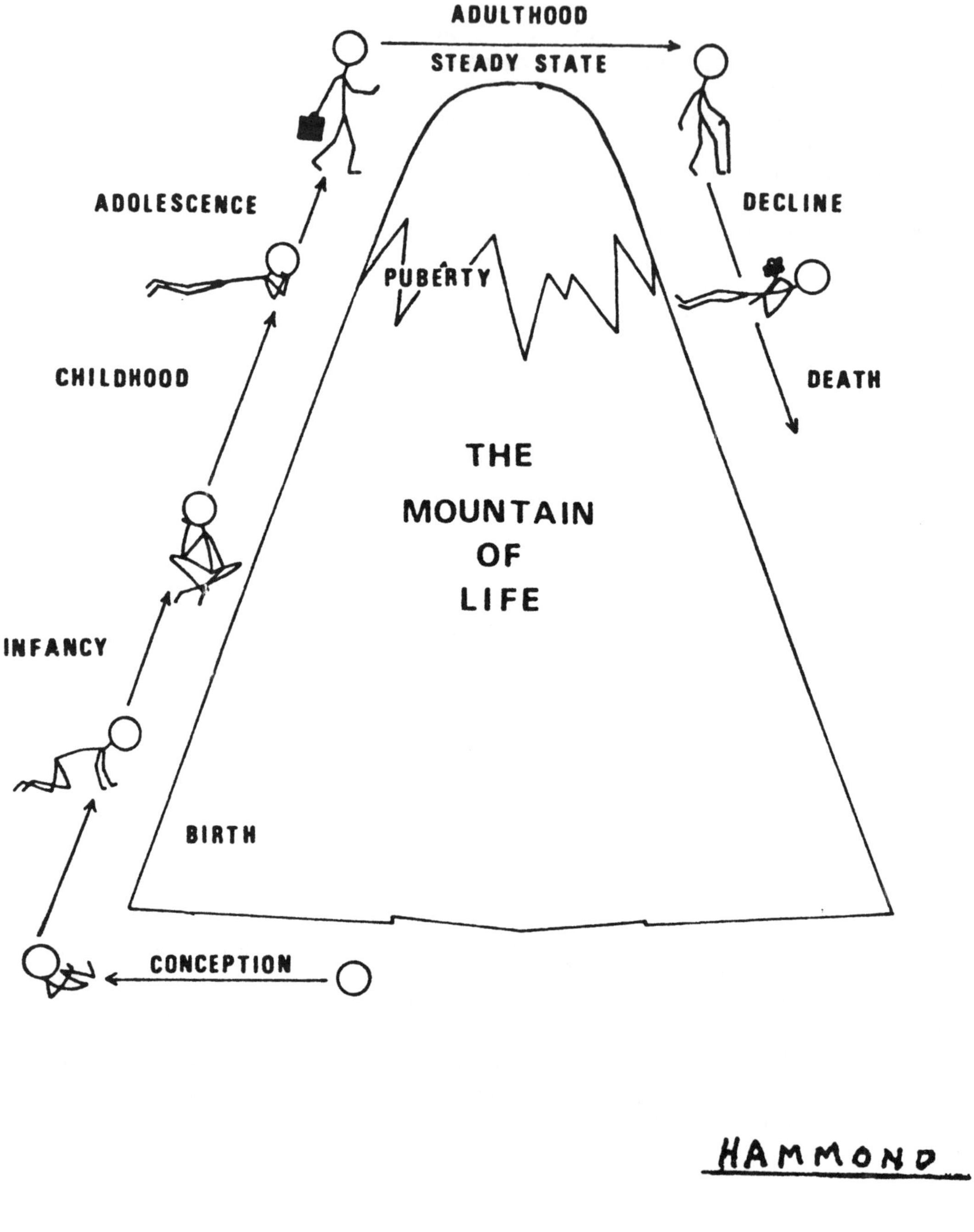
ADULTHOOD
STEADY STATE
ADOLESCENCE
DECLINE
PUBERTY
CHILDHOOD
DEATH
THE
MOUNTAIN
OF
LIFE
INFANCY
BIRTH
CONCEPTION
HAMMOND

systems. People have had to live with disorders caused by poor genetic information, poor nutrition, poor information through the senses, and, of course, being accidentally or intentionally maimed by outside forces.

However, because of this, and like all other living systems, humans have evolved elaborate defense, maintenance, and repair mechanisms to keep the system from going to disorder. The striving to survive in a state of high organization, low entropy, is built into each human system. Victims of suicide must be really desperate, depressed and unethical to themselves to deny their systems their strivings!

It is not easy for a human system to grow and go from conception to adulthood. It is difficult for a human to arrive at the end of adolescence, healthy in mind and body, and wise, having started with good genes, given good daily nutrition, and good continuous education. It is a wonder societies turn out as many good human systems as they do!

But body, brain and behavior building are not the end of the ethical considerations under the umbrella of the entropy concept. Potentially, the human system has an infinite number of behavioral options. Some are good and some are bad. That is what ethics is all about.

In the previous chapter I considered energy management, using sound (music) as an ethical deliberation. Note, the consideration was also an aesthetic one. Under the entropy concept behaviors leading to the aesthetically good, the ethically good, and the educationally good are not seen as separate phenomena. This is true for the management of matter also. So let us look at the other arts, since they do not rely solely on audio perception, but rather on visual, or some combination of the two.

The dance, gymnastics and sports are appreciated as the apex of body management. Random, uncoordinated, inefficient behavior simply will not win the hearts of observers. It is not easy, we say. And we know that only practice, patience, hard work and perseverance will lead to the peak of perfection. The same can be said for the play, the musical and the opera.

And what about writing, painting, sculpture, the crafts, architecture, drafting, building and technologizing (if that's a word)? Does not their nobility come from organizing matter, arranging matter for information, beauty and utility? These behaviors are acceptable to all cultures. They are aesthetically, ethically and educationally okay! And they are anti-entropic since they locally reduce entropy.

What is not acceptable behavior is randomly destroying information, rapidly sending energy and matter to a high state of useless entropy. That is what the destroyer does, and that is what humans do not like.

It is so easy to flow with the second law, causing energy, matter and information to reach high entropy fast. Think of the effects from an erupting volcano, a nuclear bomb exploding a city, an arsonist or bolt of lightning igniting a forest, and a murdering human or even a bee killing another human being. Sending low entropy systems to high entropy fast is very easy because it takes so little energy, matter, know-how, time or money. And that is not the acceptable way of growing and going of life on this planet. There is no reason why we should not be teaching as much to students in our schools.

Students need to know that the aesthetically and ethically good behaviors do not come easy. They take a lot of learning and a lot of labor. That is a big chunk of what worthwhile living is all about.

The growing and going human system does not do nothing. He either uses energy and matter to construct or to destroy. He can either aesthetically organize or cause ugly destruction. Education has the task of teaching how organizing energy and matter can be good and how randomizing energy and matter can be bad. The road to social order is hard, indeed. But the road to chaos, no matter how easy, is not acceptable.

For Further Reading

Robert Bruce Lindsay. 1973. *The Role of Science in Civilization,* Greenwood Press.

Alvin M. Saperstein. 1975. *Physics: Energy in the Environment,* Little, Brown and Company.

POSTSCRIPT

I mentioned that the human system has internal mechanisms to defend, maintain and repair itself, in order to live for a time in a state of open, complex, flexible, highly organized, nonequilibrium, locally reduced entropy. Without these mechanisms any human system rapidly goes to disorder, death, decay, deterioration, equilibrium and high entropy. Chaos always wins in the end. This leads authors on entropy to write whimsically:

> There is a game in town.
> Everybody must play.
> Nobody can win.

What the laws of thermodynamics and the entropy concept do not tell us is how fast a human system must lose. That is how ethics relates to time considerations in the next chapter. But for now I want to talk further about keeping matter organized outside the human system.

The internal mechanisms which defend, maintain and repair the human system have not been enough to keep most human systems healthy for very long periods of time. Thus, to further ward off death, all of the health care services have evolved. Their goal is to help the human system survive longer. And we all know how successful they have been in advanced technological societies, pushing the average life expectancy to over 40 years, then 50 and 60, and now 70.

But still accidental death and destruction remain so prevalent that insurance companies thrive on the human system's desire to ward off impending disorder, or lessen the blow. A sudden rise to high entropy is not something the human system favors. So the ethically acceptable behaviors and occupations have evolved to strive for high organization and to maintain order. Not only the arts and sciences, religions, body-mind and behavior repair, but also communications, governments, legal machinery, and indeed education are all areas that strive for social order.

When a new human system arrives at age 18, having been reared by society, and closely monitored in formal education for at least 12 of those years, and remains a rapid, high entropy producer, rather than helping society keep entropy at a minimum, then society knows they have failed. The cost of coping with randomizers is very high indeed.

It is enough that each active system, each human, each building, each machine, each environment, each highway, each bridge and each city needs constant monitoring, maintenance and repair. (With knowledge, humans have heaped an awesome responsibility on themselves—to take care of the whole planet.) There is plenty of constructive behavior to be done, without some humans causing disorder by unnecessarily killing, burning, bruising, vandalizing, littering and polluting the environment.

Again, we see that environmental ethics, health ethics, legal ethics, religious ethics and aesthetics merge as one under the entropy ethic.

TIME AND ETHICS

> Since any increase in order within the biosphere must be very small compared to the increase of entropy in the sun-earth system there is no reason to think that evolution controverts the second law of thermodynamics,...local parts of the system may...FOR A *TIME* increase in order.
>
> —Harold F. Blum
> Time's Arrow and Evolution

I have already noted that a long span of time in a healthy existence does not come without its costs. This is particularly true for living systems. Like the rest of life, the human, open system needs a continual energy flow-through to grow (build), maintain, repair and behave. We also need matter to grow (construct), maintain, and repair. And all of this growing and going from conception to a high state of organization (hopefully by adulthood) takes a great deal of time. Shall we say, 20 years, give or take a few? These three, energy, matter and time, coupled with a fourth, information (know-how), are what it takes to build an open, nonequilibrium, human system, existing in a state of locally reduced entropy.

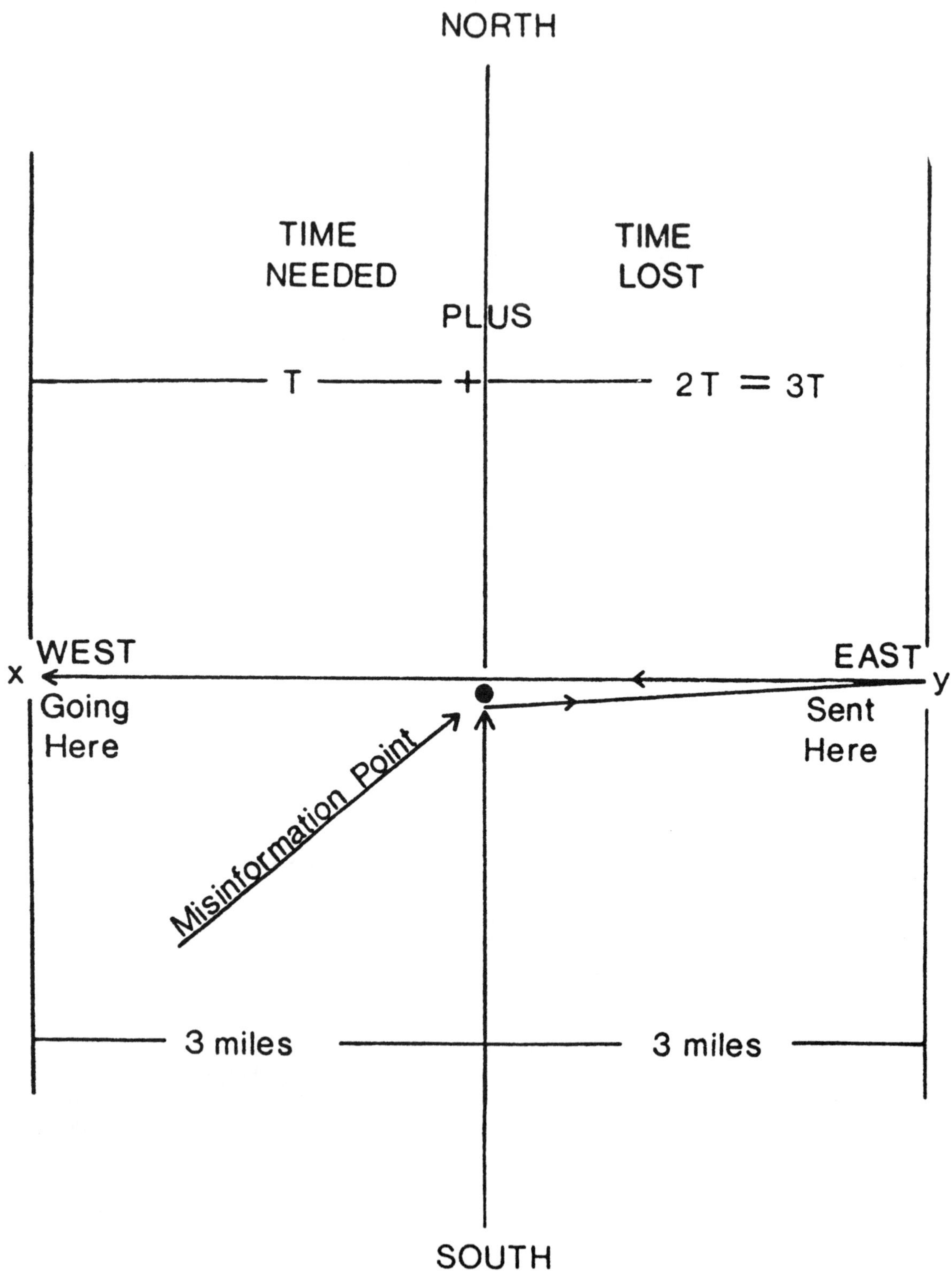
NORTH
TIME
NEEDED
PLUS
TIME
LOST
T + 2T = 3T
WEST
x
Going
Here
EAST
y
Sent
Here
Misinformation Point
3 miles
3 miles
SOUTH

Time. Do we try to save it or squander it? Time does not stand still! From the human system's point of view, time is irreversible. Since we cannot save it and it will not stand still, time is one of the accountables we must reckon with, and usually cherish. The best we can do is manage our existence within a time frame. That is really what is meant by time management.

What is the best way to manage a time frame? Be efficient and well organized! Again, energy and matter considerations are necessary. Entropy and the second law shall not be denied their roles. And know-how? As the old adage goes, plan your work and work your plan.

In other words, perform your daily chores in life in an efficient and highly organized manner. Do not waste energy and time by operating randomly. That is the time ethic! That is what is valuable to business and industry. That is how educators are taught to operate their classes. And that is how school administrators are taught to operate their schools and school systems.

Like energy conservation and using matter for constructive and creative purposes, time has its ethical component also. What I am wondering is just how many students are taught this time ethic, which is just one aspect of the total entropy ethic.

Many teachers and future teachers have asked me how to get to the entropy ethic when teaching, particularly pre-secondary learners. (More of the teaching aspects will be covered in Part 4.) As a matter of fact, many of life's unethical behaviors can be analyzed in terms of the entropy accountables.

Let us take a simple case of lying, or better, misinformation. (Withholding information can be as bad. That is called secrecy.) When misinformation or lack of information occur, then trust can break down. But that is not all. In many cases, energy, matter, money and time are caused to be wasted by lying and secrecy. The diagram at the beginning of this chapter helps explain the process. Elementary students are quick to pick up on this example.

Hypothetically place a third grade student on his bike going north, ultimately heading for town X. He dead ends at an intersection and must go right or left. He stops and asks a person which way to go, east or west. (You can also have a sign at the intersection which has been altered to give misinformation.) The third grader goes east three miles to Y, and finds he is in the wrong town. Now it takes him twice as much more time to get back to X.

Question: How much time did you lose getting to X?
Student: It cost me two times more time than I would have spent.
Question: And what about the energy you used?
Student: Ya, it cost me two times more of that, too!
Question: Your tires?
Student: Ya, they lost two times more rubber, too!
Question: That's a lot of costs?
Student: Ya, that person shouldn't 'a lied to me! He cost me plenty.

The culprit, misinformation.

Obviously, there are an endless number of scenarios which could be played out with persons of all ages. And they can be submitted to quantitative analysis. They can be studied with the help of pencil and paper, on the chalkboard, and in actuality on the school playground.

By high school the students easily relate the stories to the entropy ethic, once they know the entropy concept. Adolescents rapidly relate energy, matter and time with information. And money, too, if they are driving a car and bought the gas.

Now I want to compare this kind of reasoning about dishonesty with what was happening in the schools in the 1800's. On page 91 of William H. McGuffey's *Eclectic Third Reader* of 1847, he wrote:

> Oh, it is a dreadful thing to practice deceit. It will shut you out from heaven. Though you should escape detection as long as you live; though you should die, and your falsehood not be discovered, the time will soon come when it will be brought to light, and when the whole universe—men and angels—will be witnesses of your shame.

Some people, in and out of education, would like to go back to the McGuffey Readers as a way of getting ethical education back into the public schools. That would be fine, if the goal of formal education is to bring back the bribe of heaven and religion into the public schools.

But we are far removed from that time of the mid-1850's. We are in the silicon age of microchips, computers, robots, the accountables of energy, matter and time, systems analysis, and information management. I do not think the parents or the students want to go back 125 years to an old foundation for ethics. The students, parents and teachers I talk with tell me they feel the entropy ethic is more in tune with the times.

For Further Reading

Harold F. Blum. 1970. *Time's Arrow and Evolution,* Princeton University Press.
Ilya Prigogine. 1980. *From Being to Becoming,* W. H. Freeman and Company.

POSTSCRIPT

Time marches on! As far as the human system is concerned, time is irreversible. The human system is a dissipative structure, complex and flexible, with an uncountable number of degrees of freedom, highly organized, in a state of low entropy, regularly taking in and expelling matter, energy and information in an attempt to stay alive as long as possible.

Human systems also regulate time periods to keep their behaviors organized. There is a time to sleep, a time to work, and a time to play. We *schedule* our lives and expect the oncoming generations to learn to do the same. We value scheduling because disorder would prevail without it. This universal value of time management is a part of the entropy ethic, helping to keep a person, a family, any business or institution, any city or any state from falling into chaos.

Humans schedule meetings, production, transportation, schooling, classes and almost everything else imaginable. If you have not learned to use your time well, no organization wants you on their team.

A school, its administrators and teachers, plan and organize instruction around time periods, so that students will get information, acquire knowledge, and reach the higher levels of reasoning and psychomotor skills with the least amount of confusion or conflict, and *without wasting time.*

Teachers are encouraged to set goals, to build lesson plans, and to organize the assignments for each of their classes, and for each individual student. For teacher and student alike, attendance and promptness count. Duty to the task and respect for quality are required. And all of this planning and organizing because we value responsible human systems, which means they have learned to manage their own system's and other systems' energy, matter, information and time efficiently (not to mention money).

INFORMATION AND ETHICS

> Since stored information varies inversely with entropy, lowered entropy means a higher capacity to store information.
>
> —Lila L. Gatlin
> Information Theory and the Living System

It has been noted in Part II that there is a way to stay highly organized, in a state of low entropy, for a very long period of time, even for longer than our solar system can stay in existence. A diamond can do this, for the diamond is unchanging; rigid, not flexible; and simple, not complex. Indeed a diamond *really is forever* because its stored information exists in the orderly arrangement of matter, a state of very low entropy.

Yet, like any open, active system, each human system wears out rather rapidly. I do not think a hundred years is a very long time, given the age of the universe, the solar system, or even life on earth. If survival is to continue for any species, then genetic know-how must be passed along from one generation to the next. And this know-how for building a human system exists in the genes.

It is more than wit which leads one to say that a human system is the genes' way of perpetuating the stored information. The same can be said for any species, but as we know many species have died out. The process is not foolproof.

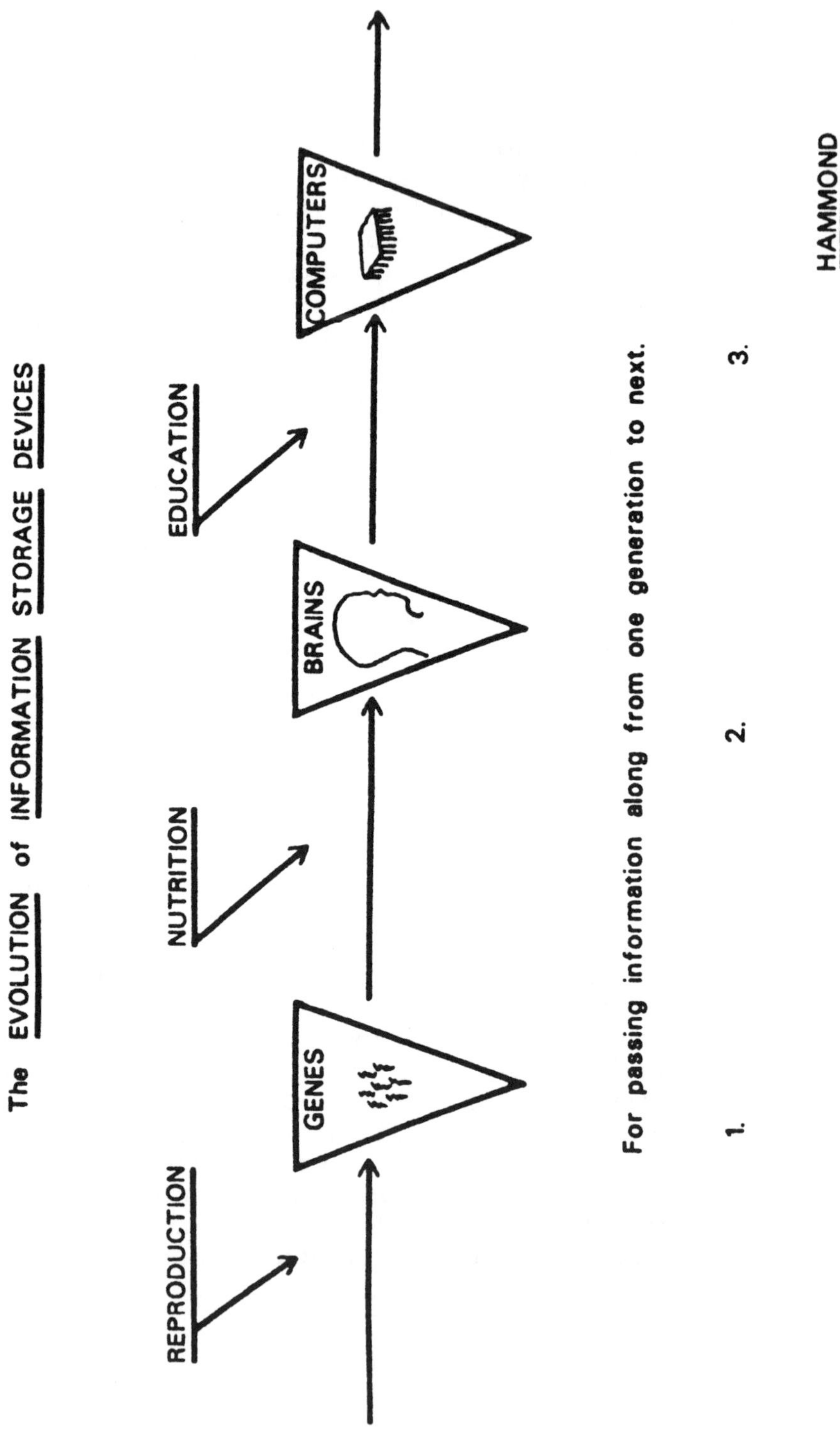
The EVOLUTION of INFORMATION STORAGE DEVICES
REPRODUCTION
NUTRITION
EDUCATION
GENES
BRAINS
COMPUTERS
For passing information along from one generation to next.
1.
2.
3.
HAMMOND

At this point in the study of the human, we do not know how much genetic information controls our behavior. My guess is with the sociobiologists, that genetic information influences more of each human system's behaviors than we suspect at this time, if not in a specific way, at least in a general sense. This is especially so when it comes to the basic needs necessary to keep the human system alive so it can perpetuate the genetic information.

The human system's matter-energy input is taken from a variety of plants and animals in various cultures. The customs and traditions surrounding eating vary from culture to culture, and indeed, from neighbor to neighbor. No two humans have identical patterns of behavior for acquiring their necessary energy-matter input. So the specifics vary. But for any individual human system, much of life's behaviors will necessarily evolve around acquiring the needed energy-matter. We are not an exception to the other living systems in this regard. What makes us different is our know-how.

As an advanced species, we no longer scavenge around randomly, hit or miss, seeking our necessary food. It has been slowly acquired know-how of orderly gathering, domesticating other animals, planting, farming and the higher technologies of food processing and transporting which allow time for other more complex brain and behavioral activities.

Knowing how to be energy efficient in organizing matter and accumulating energy is what helps give the human system more degrees of freedom.

An orderly arrangement of matter, information, know-how, knowledge, intelligence, wisdom, and thus, learning and education in their broadest sense become important to the human system. Teaching and learning know-how are valuable because they are inextricably related to the values in the three previous chapters on energy, matter and time management.

It is know-how in the genes which builds the highly organized human system. And it is know-how in the brains which makes a highly organizing human system.

The highly ordered arrangement of matter in the genes has stored enormous amounts of information and therefore exists in a state of low entropy. If we want the human to build (behave) the way the genes do, then enormous amounts of information must be acquired by each human brain, and too, it must be stored in the brain, as in the genes, in a state of low entropy. The same can be said for the systems humans build, such as libraries and computers. They store enormous amounts of information, which exist in a state of low entropy.

Thus, again, we see that misinformation is not ethically sound. We do not like it in the genes, in the brains, in the libraries, or in the computers. And, as we have seen, dishonesty can cause a waste of energy, matter, time, and yes, even money. If everyone randomly lied, know-how would not exist. What would exist would be inefficiency, random behaviors, confusion and chaos—a high state of entropy. Humans cannot tolerate these states because they are the opposite of what the genes are, and the systems they build for survival.

For Further Reading

Jeremy Campbell. 1981. *The Grammatical Man*, Simon & Schuster.
Lila L. Gatlin. 1972. *Information Theory and the Living System*, Columbia University Press.

POSTSCRIPT

In the evolution of the human system, information was, and still is, stored first in the genes. Later, additional information was, and still is, stored in the brains. Finally, more information was, and still is, stored in the objects built by humans. Knowledge, and the higher thought processes, derive from these stored information centers.

If our ethical education were in tune with our evolution, I cannot imagine anyone wanting to destroy these stored information centers. Death, by any means, destroys both genetic and brain information. Destroying paintings, libraries, computers, in fact any organized matter, destroys information. Ethical education about the value of information storage is sorely needed in our schools, since the information laws have emerged:

1. Information, genetic or learned, exists in the arrangement of matter.
2. Information, genetic or learned, regulates time, organizes matter and controls energy flow.
3. Information, genetic or learned, is inversely related to entropy.
4. Information, genetic or learned, is created and destroyed.

Without understanding these aspects of information, we cannot truly know ourselves or any other human system.

Lila Gatlin said:

> The second law of thermodynamics is indeed an order-degrading principle *in itself and without constraint;* but when we place it under the control of the higher law of information theory, it becomes directly responsible for the production of order of a very important type. This is why life has arisen.

Just as energy, matter and time are basic to the entropy ethic, so too is an understanding of information and its relationship to entropy.

THE HUMAN SYSTEM AND ETHICS

> ... science may soon be in a position to investigate the very origin and meaning of human values, from which all ethical pronouncements and much of political practice flow.
>
> Philosophers themselves, most of whom lack an evolutionary perspective, have not devoted much time to the problem. They examine the precepts of ethical systems with reference to their consequences and not their origins.
>
> —Edward O. Wilson
> On Human Nature

Albert Einstein must have had similar thoughts to Wilson's. Einstein seemed to sense the possibility of using scientific findings as guidelines for human conduct. In his book, *Out of My Later Years,* he has a short, two page essay on, "The Laws of Science and the Laws of Ethics," in which he wrote, "...ethical directives can be made rational and coherent by logical thinking and empirical knowledge."

Our origins? What is the evolution through time of a human system? What is life all about anyway? I ask these questions because it seems to me we should at least place some value on knowledge about how we got to be what we are, and on what we have come to be. If there is not *some value there,* then we must be totally worthless.

If politicians, pedagogues, preachers, publishers, psychologists and philosophers could agree on our origins and what we are, then maybe, just maybe, the schools could get on with some relevant ethical education.

Thus far in this book, I have tried to show that we human systems are energy, matter, information and time using, irreversible, open systems. The question is, "How do we get from what we are doing to what we should be doing?" I mention this because it is a question which bothers many philosophers and educators. But when I talk to teenagers about this question, it does not bother them a bit. Sometimes I think it is just a matter of older people set in their ways of thinking, while adolescents are more open to new ideas.

After all, parents and teachers are constantly telling each new human system what to do as he goes from birth to adulthood. Once they know what they are, it is not difficult for teenagers to figure out why they ought to behave one way rather than another. On the other hand, few people can tell a philosopher or an educator what they are, let alone how they ought to behave.

But, behave we all do, and some are good and some are bad. So we need to look carefully at what is meant by ethics and human behavior.

Moral philosophy and ethics are synonymous. They are concerned with right and wrong, good and bad, and what should be done to make a better life for humans.

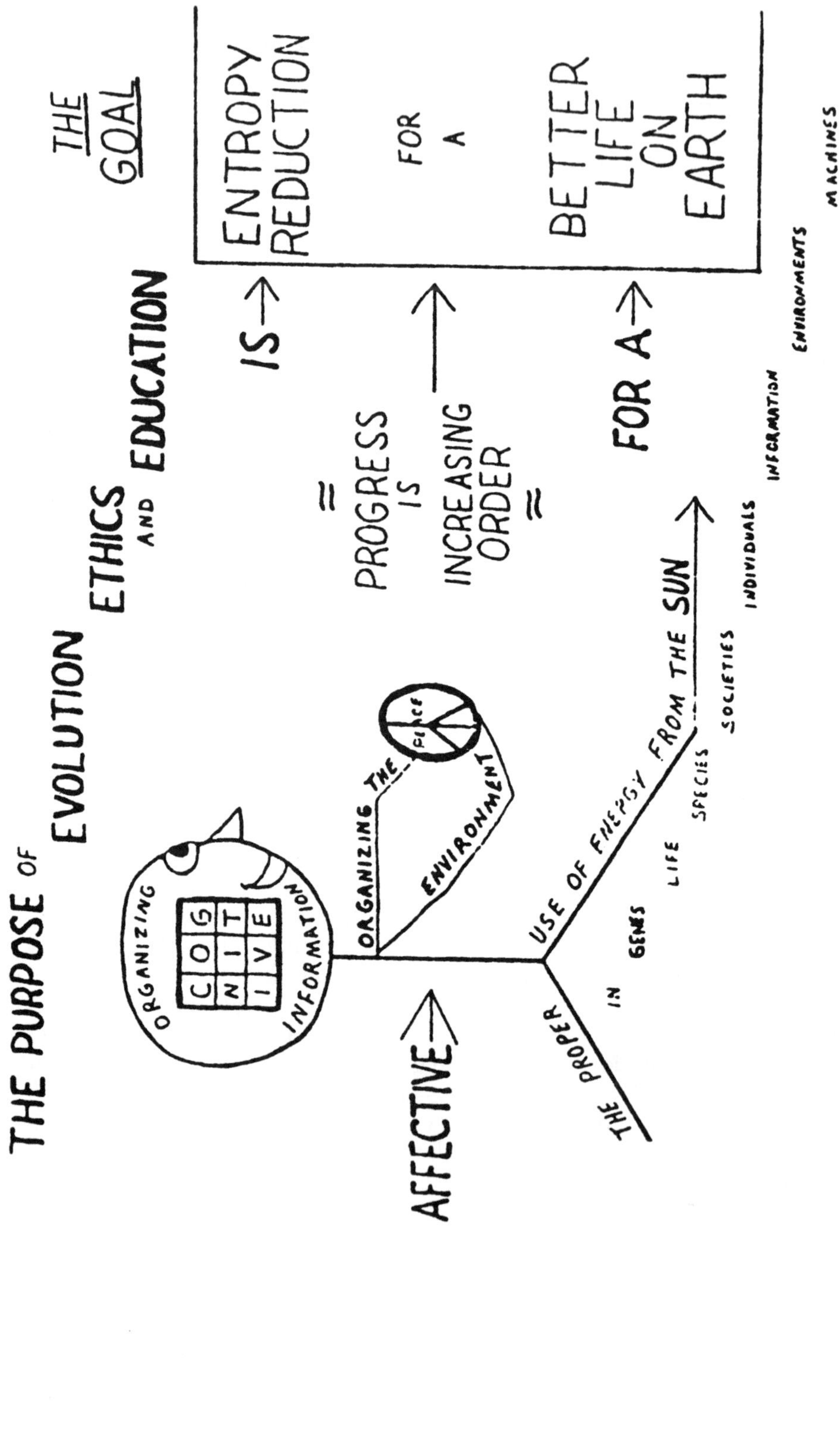
THE PURPOSE OF
EVOLUTION
ETHICS
AND
EDUCATION
IS →
THE GOAL
ENTROPY REDUCTION
FOR A
BETTER LIFE ON EARTH
=
PROGRESS IS
INCREASING ORDER
=
FOR A →
AFFECTIVE →
ORGANIZING
COG NIT IVE
INFORMATION
ORGANIZING THE PEACE
ENVIRONMENT
THE PROPER
USE OF ENERGY FROM THE SUN
IN
GENES
LIFE
SPECIES
SOCIETIES
INDIVIDUALS
INFORMATION
ENVIRONMENTS
MACHINES

Ethical instruction and moral education are the same. They are concerned with the teaching of right and wrong and good and bad behaviors. They are also concerned with helping students to consider what should be done to make a better life for humans.

Behavior refers to what humans do or say as they interact with their environment. Behavior also means using energy. To ask: "How should I behave?"; "What should I do?"; or "How should I conduct myself?"; is the same as asking, "How should I use my energy?"

Individuals move by using energy. Information is carried to an individual via energy. Energy flows in various forms from various sources. Humans have learned to tap the sources of energy and to control its flow. With the use of energy an individual can, therefore, create or destroy; organize or randomize; be more efficient or less efficient; cause more order or disorder; be constructive or destructive; conserve information, matter, and energy, or dissipate them quickly and irretrievably.

Humans can very easily produce high entropy—a rapid dissipation of information, matter, and energy to randomness—as can a bolt of lightning, a match, or a spark from flint. Very little time, money, information, matter, or energy is needed to take the randomizing route with nature's second law. However, the constructive task of reducing entropy is always more difficult, because a great deal of time, information, matter, and energy (not to mention money) are needed to build the better life in harmony with nature's laws.

The Term Entropy Reduction

Before looking at some guidelines for human behavior which are inherent in the total entropy concept, it is necessary to analyze the term entropy reduction itself. Necessary, because looking back through the literature on the entropy concept, it can be noted that many authors use different terms to indicate the movement away from randomness and thus the irretrievable loss of energy and information.

Because of the newness of the concept, no generally accepted term is used in the current literature. So the term entropy reduction is used here with the same meaning as all of the following. It is said that life: plants, animals, humans, and their behaviors, to be successful as a part of nature, must:

> decrease the amount of entropy production,
> minimize the amount of entropy production,
> thrive on negative entropy or negentropy,
> retard entropy,
> absorb entropy,
> diminish entropy,
> consume entropy,
> decrease entropy,
> lessen entropy,

cut down entropy,
lower entropy,
curtail entropy,
be anti-entropic,
stop wasting free energy,

and humans should have an entropy ethic so they will know why they are taught to do what they are doing. Along with the term entropy reduction, all of the above refer to reducing the speed at which energy, matter, and information are irretrievably lost or rendered unusable or have to be replaced.

Entropy reduction refers to how humans use energy as open systems to manipulate other energy, matter, and thus information outside themselves. As will be noted, the total concept has far-reaching ethical implications.

Some Guidelines for Ethical Human Behavior

What does it mean to consume entropy, or to minimize entropy production? What does it mean to unnecessarily create more entropy? How can humans decrease entropy? In other words, what kinds of human behaviors are necessary to reduce entropy? And, what kinds of human behaviors degrade life by producing entropy? To answer these questions, or arrive at guidelines for ethical human behavior, entropy-reducing behaviors must be compared to entropy producing behaviors.

1. SHOULD HUMANS *CREATE* AND NURTURE, OR SHOULD THEY *DESTROY* AND DISSIPATE?

An obvious obligation for survival is to continue creating and nurturing life on earth. Destruction of life is an increase in entropy. Nurturing plants and animals reduces entropy.

A person who unnecessarily destroys life by any means becomes an entropy producer. Therefore it is logical that societies pass laws against the arsonist who burns down the forests and the murderer who destroys other life forms.

A more contemporary problem is the over creation of life. An agriculturist would not think of overpopulating an acre of land, cramming in any kind and any number of plants, until chaos is reached in the striving for energy, matter, and space. Humans have not always taken these entities into account in the development of their own living centers.

But it is clearly entropy producing to overpopulate a corn field, a canary cage, a country, a city, or a highway, just as it is entropy producing to destroy. Entropy reduction always calls for the proper amount and kind of energy, matter, and information input into a system in order that the system can function efficiently and creatively and progress in further entropy reduction.

The nurturing of an animate system is a delicate and time consuming enterprise—especially a human system. This entropy reducing activity takes not only time, but a lot of energy, matter, information, and money. By comparison, the act of destroying a living system can be accomplished with very little expenditure of the money, information, matter, energy, or time accountables. Destruction, or entropy production, is very easy.

Hardly any of the accountables are needed for destruction. Take, for example, an act of vandalism, or the destruction of inanimate objects. This is clearly an entropy producing behavior which plagues schools and society without intermission. Think of the amount of the accountables necessary to create the schools, subway stations, and all other structures of a society. Then think of the almost total lack of the accountables needed to vandalize the matter which was put together in the entropy reducing process. And, following the vandalism, if the structures are to be restored, then even more of the accountables must be used. Is it any wonder that the crimes of vandalism rate high on societies' lists of behaviors to be corrected by ethical instruction? Vandalism, murder, and destruction in general are clearly unethical behaviors. They are wrong behaviors when the concept of entropy reduction is understood.

Humans, like all other animals, plants, and organized systems on earth, have had to cope with the destructive forces of wind and rain and snow and floods. People have had to repair their structures after the effects of hurricanes, typhoons, tornadoes, dust devils, tsunamis, tidal waves, earthquakes, avalanches, volcanoes, lightning, and fires; not to mention the debilitating work of lice, rats, mice, termites, worms, mosquitoes, fungi, bacteria, viruses, poisons, rusts, and other sources of decay and deterioration.

Destroying entropy reducing systems will always be an easy task. Creating new and better entropy reducing systems is necessary for the survival of life on earth. It would seem that the ethical instruction of the young must teach students that their role in life is to create, and to help maintain entropy reducing systems.

2. SHOULD HUMANS *ORGANIZE* OR *RANDOMIZE* THEIR ENVIRONMENT?

When matter is organized, entropy is reduced and information is made available to the human senses. When matter is being dispersed more and more randomly, then information is being made less and less available to the human senses.

In many advanced nations the acts of littering and polluting the atmosphere, hydrosphere, and lithosphere have only recently become criminal offenses. To beautify the environment has often been the reason used to convince the public they should stop the entropy producing behavior of distributing matter randomly.

But there is more to an analysis of this domain than the aesthetic value. Littering, like all entropy producing behaviors, takes hardly any information, time, money, matter, or energy. Keeping matter organized always takes more of these accountables.

This applies not only to the highways and land in general, but to the air and water as well as the cities, campuses, libraries, factories, businesses, schools, hospitals, homes, rooms, and on and on. And, again, after littering takes place even more of the accountables are needed to reorganize the matter.

The organs inside a living system can only function adequately and efficiently by highly organized matter, and thus entropy reduction occurs. The organizations built by and outside living systems can also only function adequately and efficiently by a reduction of their entropy.

Randomizing the environment is clearly wrong behavior when the entropy reducing concept is understood. Students can be given adequate reasons why they should put their toys and tools away, tidy their rooms, help control the trash, not litter hospital halls, school grounds, businesses, and the great outdoors. Recognizing the importance of keeping matter organized, and the obligation to organize, and why society must demand such behavior, are all implications of entropy reduction for ethical instruction.

Randomizing entropy reducing systems will always be an easy task. Organizing new and better entropy reducing systems is an obligation of life on earth. Therefore, the ethical instruction of the young should teach students to organize and to help arrange matter to further entropy reduction.

3. IS *EFFICIENCY* MORE IMPORTANT THAN *INEFFICIENCY*?

When humans cause matter and energy, including themselves, to flow along specific routes, they are attempting to minimize entropy production. Efficiency is a measure of their success. Unaware of the entropy concept, and the need to reduce entropy for survival, humans have nevertheless been very successful in efficient energy utilization. The internal structuring of a human was (and is) guided by the information passed on in the genes. On the other hand, behavioral efficiency may have come about by an early association with time saving, and later, with saving money. Regardless of why technology advanced, humans developed tools, simple machines, categorized systems, streets, electric circuits, complex machines, waterways, highways, skyways, and hallways. All of the applied sciences are testimony to the attempt to build more and more efficiency into human made systems.

The ants, worms, fish, and birds have their efficient routes too; however, no other animal on earth has the ability to rapidly reduce more entropy outside themselves than humans.

Getting more efficiency out of a system means building the system so that the energy input is not so rapidly wasted. In the case of building machines, inefficiency means losing the energy to unnecessary friction. IT IS ENTROPY REDUCING TO SLOW DOWN THE RATE OF ENERGY DISSIPATION TO AN UNUSEFUL STATE. Though it is not a matter of creating or destroying energy, in the above sense, energy can be thought of as being conserved. Unnecessarily dissipating energy to an unusable state is called wasting energy.

Therefore, energy is wasted by burning a forest or a library, by inadequate insulation, and by building unneeded friction into a drill press or an automobile.

It is hardly acceptable behavior to build automobiles with less than round wheels, loosely attached pipes and electrical attachments, poorly fitting engine components, and other highly abrasive moving parts. The consequences would be obvious. The energy in the chemically bonded gasoline would be less transformed into the motion of the total car system, and more transformed into the rapidly lost heat, light, and sound energies.

All of the applied sciences have the task of making the systems they work with more efficient, and thus, less entropy producing. From engineering to architecture, and including all the health services, the purpose is to build less friction, conflict, confusion, and randomness into their systems. Rather, they strive to construct and maintain their systems in the most efficient state possible.

4. IS *ORDER* VALUED MORE THAN *DISORDER*?

When calorimetry, thermodynamics, and eventually, entropy considerations were in their embryonic stages of development the concept of order versus disorder was very simply confined by the physicist to isolated and closed systems. The sociologists talked of order versus disorder in societies too, but since the mathematics of the physical scientists could not be applied by people working with human behavior, there seemed to be no way to develop a foundation for ethics based on natural phenomena. As is noted in the appendix, that stalemate was broken by the introduction of information theory and open systems. To show more clearly the link between molecular motion, information, and open systems, a detailed analysis of order versus disorder follows.

Given: A small, solid, dry ice replica of a dog is in a hypothetically large, isolated chamber (remembering that isolated means no exchange of matter or energy with the universe outside the isolated system). The moving molecules in the room are at fifty degrees Celsius, and the dog is at freezing point of carbon dioxide. The temperature in the room is determined by the motion of the air molecules. The air molecules are in a state of high disorder because they are randomly distributed throughout the room.

The dog is in a state of low entropy because its crystalline, solidly formed molecules contain less motion, therefore less temperature, and thereby the dog's molecules are more ordered and more concentrated. Leaving the system (cold dog in hot room) to itself, the disordered, rapidly and randomly moving air molecules bombard the orderly dog. The dog's carbon dioxide molecules take on energy from the air molecules and swiftly sublimate. What was once the form of a dog soon loses its shape. The dog's molecules are randomized throughout the chamber.

The temperature becomes the same throughout the system as equilibrium is reached. The air molecules move more slowly than they did originally. Their entropy has decreased. The dog's molecules are moving more rapidly than they were originally. The dog's entropy has increased. The overall entropy inside the chamber has also increased. The total entropy inside an isolated system has a one directional flow. The disorder always becomes greater.

If the system described had been a closed, not isolated system, more energy could have been put into the system as the dog's molecules were dissipating outward. Then the dog would have lost the orderly arrangement of its molecules much faster. In other words, its entropy would have increased much faster, because of the more forceful and the more frequent bombardments on the dog by the air molecules.

In either the isolated system, or a closed system with energy passing in and out, the probability of the dog returning to its original ordered shape, temperature, and entropy state is infinitely small. The dog is lost to disorder, never to be seen again in an ordered form.

This is precisely the point where classical thermodynamics stopped and differs from the contemporary view. In contemporary thermodynamic reasoning, the loss in the arrangement of the dog's molecules is also a loss in information. As the dog's molecules went to a greater entropy, its shape vanished, and information about the dog was lost. High information (an orderly arrangement of matter) means reduced entropy (also an orderly arrangement of matter).

Humans strive to arrange matter in orderly fashions. They can, and they do, because they are not isolated or closed systems. Rather, they are open systems, exchanging both energy and matter with their environment. Were a team of mechanical and chemical engineers asked to go into the chamber and return the dog to its original shape, temperature, and entropy state, they could. They would separate the carbon dioxide molecules, build a mold, put in the carbon dioxide, and lower its temperature. The dog could be seen again, yielding information, and the dog's entropy would be reduced again. Of course the expense of the success of reducing entropy, as always, is in terms of the additional expenditure of time, money, matter, and energy. And further, though the dog's molecules would not occupy the same spaces they did originally, what gives humans this certain measure of success, is guiding information on the use of matter and energy.

Such analyses have many implications for physical, biological, and social science education, curriculum development, and for the philosophy of education in general. By analyzing ordering processes versus disordering processes, or entropy reduction versus entropy production, many guidelines for human behavior can be generated. Also, by using human behavioral examples, thermodynamics may be better understood.

5. SHOULD HUMANS BE *CONSTRUCTIVE* OR *DESTRUCTIVE*?

Using thermodynamic examples, life and living may be better understood also. When the genetic information from the male and the female are united and a new living system is conceived, controlled construction begins for a new entropy reducing system (zygote). Being constructive or constructed is the first task performed by the newly developing, living, open system (embryo).

Physical chemistry, molecular biology, developmental anatomy, and genetics show how the new living system takes in and releases matter and energy, and thereby constructs a new individual. Apparently the genetic information controls not only the construction, but also, for many species, the inherited characteristics help determine the behavior of the individual organisms after birth. Humans, however, have a long way to grow and go after they are born. After birth their internal construction continues until the individual dies. The genetic information, cell parts, and complete cells are constructed over and over again, approximately a billion times each day. Entropy reduction inside the living system is thus maintained. But the behavior of a human system is not so well delineated toward constructive ends.

Surveying the disciplines of child psychology and child development, some important clues to human values can be reached. Obviously, infants cannot construct when they are born. Even the most unknowing parents soon realize that their newborn is helpless, uncoordinated, and has only the rudiments of psycho-motor skills. During the first few years an infant's behavior is far from constructive. In fact, entropy production (causing disorder) is more apt to be displayed than entropy reduction (bringing about order in the environment). This leads to the conclusion that the genes have less influence in the domain of human behavior than in other species. It also emphasizes the need for a specific kind of education and ethics, which will yield valid guidelines for human behavior.

Whether the genes have a strong or a weak influence in bringing about the learning of constructive behaviors is not relevant here. Children can go from destructive to constructive behaviors, from uncoordinated to coordinated behaviors, and from being little entropy producers to entropy reducers—if they have learned the right behaviors.

Infants cannot stack the blocks, they can only knock them over. They cannot organize the blocks, or order them, or construct anything with them. They can only randomize the blocks, destroy any structures, and send matter into disarray.

Children are at first unavoidably destructive in their behaving. The example of the blocks could be expanded to food, toys, puzzles, clothes, sand, dirt, and many other items which children try to manipulate. Children randomize the environment, because they lack psycho-motor skills. One of the primary tasks of parents, schools, and society is to replace unlearned destructive behaviors with learned constructive behaviors.

So much value is placed on constructive behaviors, society can hardly tolerate the consequences if the task is not accomplished. Destructive and disruptive behaviors become discipline problems in the schools. Violence and vandalism are not condoned, and individuals committing crimes of destruction are finally ostracized from society.

Parents, educators, and society make demands for construction as though they understood the whole entropy concept. Parents punish randomizing behavior, schools have rules against destructive behavior, and societies pass laws against behavior which produces entropy. Entropy reduction (construction) is necessary for the internal growth of an individual, and it is also valued, and necessary in the behavior among individuals. Should the value system swing in favor of entropy production (destruction) then...?

There appears to be no alternative to the fact that life on earth is wedded to entropy reduction. Without the capability to reduce entropy, living systems would not exist. Therefore it seems mandatory that this knowledge must be incorporated into the philosophy of education, and the ethical instruction of the young.

For Further Reading

Edward O. Wilson. 1978. *On Human Nature*, Harvard University Press.
Morton Hunt. 1982. *The Universe Within*, Simon and Schuster.

POSTSCRIPT

The entropy ethic shows that guidelines for human behavior can be generated from what is known about natural phenomena. Tying the disciplines of science together, and looking at life and living holistically, the concept of entropy reduction offers ethical directives with many implications.

By the use of energy, matter, and information, life is found to go in the opposite direction of normal entropy flow. Research proves that matter and energy tend to become randomized, disordered, dissipated, and therefore less arranged and less usable. Yet life organizes and arranges matter and energy, creating order out of chaos and constructing as efficiently as possible.

If survival and a peaceful existence are to be maintained on earth, some ethical directives become quite clear from an analysis of entropy reduction. It is a part of the human system, as a part of life, to try to conserve–not waste–energy, matter, and information. Killing, murder, and arson rapidly send matter, energy and information into a chaotic state, and are not acceptable behaviors. Energy, matter, and space shortages are caused by overpopulation, which leads to more chaos. Stealing, suicide, and other forms of self and structure mutilation cannot be sanctioned from an analysis of entropy reduction. Violence, vandalism, and other disruptive behaviors which throw energy, matter, and information into disarray, cannot be condoned. Polluting and littering cause any environment to become disheveled, and cost even more energy in the clean up. These behaviors are as wasteful as ineffiency.

There are also many obvious implications which should be mentioned without a detailed deduction from the theory of entropy reduction. Poorly insulated single dwelling homes and other structures allow energy to be rapidly and irretrievably dissipated. Too many people driving too many big cars too fast, wastes energy. Fighting and war and the production of destructive weapons have consumed countless amounts of energy and matter, not to mention time and money. Misinformation and the lack of information do the same thing. Lying and secrecy often cause other people to use the accountables unnecessarily.

On the other hand, most of life's behaviors and occupations strive for order and efficiency in the conservation of energy and matter. Just as there must be freedom to investigate and to learn, so too, there is an obligation to order and to learn to order.

The multitude of psycho-motor skills learned by students in the schools are testimony to the importance of the ordering process necessary for entropy reduction. If the theory of entropy reduction were to be used for ethical instruction, the implication is that students could realize the importance of the skills they are being taught. At the same time they would be learning some guidelines for human behavior.

Some Advantages to Teaching Entropy Reduction

The first important advantage then, is that students could understand why they are learning what they are learning. If they learned the values of organizing matter, less rapid energy dissipation, arranging, categorizing, and taxonomizing, students might better realize the importance of what they are being taught.

Students might also become more aware of their role as a living, learning system in nature. If the values of opposing vandalism, violence, and other destructive behaviors are to be adopted by students, then students need to know they are a part of that larger system—nature. The concept of entropy reduction calls for a holistic, and an interdisciplinary approach to teaching about natural phenomena, with the student firmly integrated within the system.

Presently the concepts about natural phenomena are scattered throughout a student's school years. Also, the secondary courses in science tend to be specialized, still following the more narrow limits of the disciplines delineated by university departments. Students rarely get the opportunity to put together the pieces of the puzzle of natural phenomena. Rather, the pieces are almost randomly thrown to students, as inquiry, self-discovery, and the scientific method are emphasized. The holistic and interdisciplinary approach of entropy reduction calls for helping students understand the conflict between order and disorder throughout nature, which finally culminate in an awareness of a living system's obligation to order. Such an awareness is not readily obtainable in the random approach to learning about nature, as found in the schools today.

A further advantage to emphasizing entropy and entropy reduction concepts is that mathematics need not be a delimiting factor. At present the mathematical requirements keep many students from studying energy and matter, while information transmission is virtually ignored. By looking at natural phenomena from the organizing versus the randomizing point of view, the door remains open for all students to participate, regardless of their background. Many students who are currently left behind could acquire some concepts about astronomy, energy flow, the bonding of matter, genetics, information theory, and other areas of which they are left ignorant today.

The pure and applied arts and sciences have always been, and still are, highly entropy reducing human enterprises. The concepts contained in a course which emphasizes creating versus destroying would certainly be beneficial to the future generations' occupations, whether the students choose merchandising or music, teaching or technology.

Finally, teaching entropy reduction strikes directly at the bad behaviors which are burdening schools and societies. The concerns about vandalism and all other forms of destructive behavior could be tackled head on. And, being based in a knowledge of natural phenomena, entropy reduction is not sexist, racist, or nationalistic. It does not depend on the world's religions, or any other philosophical format. Thus an entropy to ethics course could be fully implemented into contemporary societies' school systems any place on the planet, imparting an unwavering set of values which would last a long, long time into the future, and must if humans are to survive.

PART 4

Educating the Human System

1. *Introduction*
2. *Teaching About Time*
3. *Teaching About Information*
4. *Teaching About Matter*
5. *Teaching About Energy*
6. *Teaching the Human System*

INTRODUCTION

Energy-matter is in constant motion in our universe. Energy-matter changes, one into the other. Energy-matter becomes concentrated, builds up to a state of high organization, exists for a time, then falls apart and disperses. The binding forces in nature can be as transient as the short existence of a single snowflake exposed to the sun or as long lasting as a perfect diamond.

Using the energy from the sun and the matter from the planet earth, each human system begins, grows by self-organization, then goes to disorder, deterioration, death, and ultimately dispersion. Survival is a very transient state for every human system, just as it is for every member of every living species on our planet.

Since long before recorded history, humans have dreamed away the end of survival—death. The building of beliefs about gods and demons, spirits and souls, heavens and hells, and reincarnations and embalmings are all testimony to the human desire to bypass death. The dilemma of survival (the human system rising briefly to self-organization and self-awareness only to be stricken down by disorder and death) became interwoven with sin, and thus, ethical considerations. These are the foundations for ethics which we cannot put back into the public schools because of the separation of church and state in our pluralistic society.

This linkage of ethics to religions is also what has kept most educators from getting involved in research leading to better ethical education for our oncoming generations. As long as religions claimed the only approach to ethical considerations educators were hesitant to begin research in ethical education, and most still are. Somehow that deadlock must be broken if we educators are to help build a better life for humans on this planet.

It seems to me that what we educators need to do is simply steer clear of all belief systems about death and what might follow. It would be better for educational research to dwell on survival, what it takes for "good" survival, and leave the belief business to religions.

Again, I am back to asking the basic question, "What is the trick to the survival of a species, any species, but particularly the human species?" That is what we must ask ourselves when looking for a foundation for ethics; how to go about teaching a foundation for ethics follows therefrom.

My work with the entropy concept has given me some specific guidelines for approaching the ethical components with students in early childhood through elementary, secondary, undergraduate and graduate programs and beyond. The process is neither complex nor lengthy. Yet, I have found it one of the most satisfying of my teaching career. Probably because the feedback from my students has been so positive.

The center of my teaching is always the student. The student determines who we teach, what we teach and how we teach. The student is the one who will have to hold the values of which ideas are important and which ideas are not important. The student is the one who will need the guidelines regarding which behaviors are good and which behaviors are bad, and why. The student is the one who will have to behave from those ideas and guidelines.

If there is one characteristic of the entropy ethic I should applaud the loudest, it is the fact that there is no need to teach or preach the ethics that are involved in the total entropy concept per se. All the teacher need do is arrange the course content in such a way that it leads to ethical considerations. The students decide which are right and which are wrong behaviors, and why.

To try preaching or teaching every behavior as good or bad would be impossible, since humans can perform an infinite variety of behaviors. That route to ethical education is not practical and would probably be counterproductive, since many behaviors are not worth the time or trouble of contemplating their goodness or badness.

What we can do is teach the basics of what we are and how we came to be what we are; and how and why our personal education puts us through the rigorous behavioral controls it does to make us good managers of time, energy and matter, and the information and systems that can be derived therefrom.

The brain of the human system has evolved to be the most flexible, three pound, information storage device on earth. The history of the world's cultures shows us that any language and any belief can be inculcated in the human system. The task of education is to decide what to teach the human system, and how best to teach it.

If education were to decide to openly teach oncoming generations the values involved in bringing order out of chaos, then some guidelines for educators would become quite clear. Thus far, in this book, the accountables of time, information, matter and energy have been needed to explain the entropy ethic. The management and manipulation of these four entities are needed to bring order out of chaos. Fortunately, these four accountables are easy for students to remember since their acronym is *T.I.M.E.*

When the hypothetical Maxwell's Demon was to bring more organization to a state of randomly distributed, gaseous molecules, these were the four accountables needed to perform the task. Reducing entropy always requires taking time, information, matter and energy into account.

Though Maxwell's Demon presented a paradox to scientists for decades, today, with hindsight and the help of information theory and systems analysis, it has become easy to liken the creature (which is what Maxwell called it) to a ship's captain, a bank president, an orchestra conductor, a school principal, or a classroom teacher.

The schools are the only places in our society where each developing human system is required to go. Consequently, society has come to use the required educational environment to help thwart and/or correct many of society's and individual disorders.

Science has found that having good genetic information, having good nutrition, having a good early environment, and having good receptors hooked up to a good brain are all necessary ingredients for a good education to take place.

Thus, we have seen the schools become partners with society in maintaining the health of our society. Biological, brain and behavioral disorders are expected to be curbed and/or coped with and cured in the schools.

The schools are to make sure that each entering child has the necessary vaccinations to ward off contagious diseases. The schools should check for genetic and mental disorders. The schools must test for any seeing, hearing, speaking, and writing disorders. The skin, scalp, teeth, heart, hands, and temperature are to be monitored at a moment's notice. Required medications are to be kept on hand. A full time nurse is always to be available. The children are to be properly fed each day and the environment is to be clean and neat with proper lighting, noise and temperature controls. The schools are to make sure each child is free from any hindrances which might curtail learning.

The schools are to practice fire drills, earthquake drills, flood drills, hurricane drills and tornado drills. And the schools must get each child to and from school safely.

Now, let the education begin!

Upon entering school, children have only barely learned to feed and clothe themselves. Coordination is in its infancy. Many psychomotor skills are yet to be learned. This learning of bodily self control will progress for at least ten years for each new human system, to the end of high school for most humans, on into college for many students for whom the requirement will not cease until adulthood.

For about fifteen years of each new, human system's life, the schools will educate for good, acceptable behaviors. The schools are to encourage creative and constructive behaviors. The schools are to discourage destructive behaviors.

The schools are to build new and efficient organizers, not inefficient, uncoordinated randomizers. Certainly, it would be easier and cost less money to lead each new human system down the road to chaotic, destructive behaviors. But society has decided the school's role is to help bring order out of chaos, and that is a very difficult and costly goal, indeed. Working to bring order out of chaos in the psychomotor development of the human system is very expensive. The time, information, matter and energy accountables must be very valuable to educators.

Upon entering school, children have only barely learned to state their feelings. Learning to analyze, synthesize and evaluate their own emotions will take each human system years to accomplish. The schools are expected to help.

Society wants the schools to produce new, good human citizens who have learned to control their emotions. The schools are to counsel, guide, and test each student toward a realistic, self awareness. Educating to produce mental disorders is not allowed.

At the slightest hint of a mental disturbance, the school is to have the where-with-all to muster forth the principal, parents, teacher, counselor, school psychologist, and special education specialist to help keep each child from sinking into despair, depression, or other affective calamities. Emotions which disrupt the education of the human system must somehow be soothed.

Upon entering school, children have only barely learned a few rudiments of knowledge. Cognitive development has a long, long way to go and grow inside each human system's brain. The schools are expected to help build the knowledge base, develop critical, creative and accurate thinking, and finally, produce a logical, abstract reasoner; one who can solve complex problems by logically evaluating issues from many points of view.

All of this educating for good functioning brains and behaviors comes with a very high price tag. The goals society has set for the schools are inversely related to entropy. Therefore, the monetary costs will be high because time, information, matter and energy will have to be used and accounted for. There is no other way to bring order out of chaos in nature, and the students and the schools are a part of nature.

The parents in the society want their new, human systems to be as free as possible from disorders in their biological, brain, and behavioral developments. Society's contract with the schools and their personnel is to make that dream come true.

TEACHING ABOUT TIME

The T.I.M.E. acronym is filled with values, as we will see in the next four chapters. When an acronym is worthwhile it ought to tell us a lot. These chapters will summarize some of the values we should be teaching oncoming generations in the public schools. It would be nice if the values were taught in the homes, before children started to school, but unfortunately that is not always the case.

In the T.I.M.E. acronym the "T" represents "time." That is easy to remember, and time is very valuable. Since it always takes time to build, construct or organize any system, a good time manager is very valuable. Also, not letting a system go to disorder means time will have to be spent. It does not matter whether we are talking about the four dimensional, open system above an airport (where we really would like to have zero entropy), the classroom, a car, the brain, or simply a set of teeth. Much time will have to be spent on any active system to keep the system from going to disorder. Time is an accountable and a value we all must learn.

We say, "Don't waste time!" We do "Time on Task" experiments to show what we already know. Skills and knowledge will be learned quicker and stronger if we spend more time on them.

Time management seminars are very popular, not only for teachers and school administrators, but also in occupations such as industry, business, finance, health, government and, of course, the military, to mention only a few.

I cannot imagine an institution of higher education, an employer or any group of parents who would not value a high school diploma which stated that a student had mastered the knowledge and skills to be an excellent manager of time. This first component of the T.I.M.E. acronym must be very important, necessary and worthwhile, and therefore a value to be taken into account by educators.

It is probably the case that time means nothing to the plants and other animals on this planet. The genes live on from one generation to the next without having to think about or "remember" the past (in a brain sense and not in a genetic sense). Nor do they have to think about or "anticipate" the future. The civilized, educated human lives a quite different existence, studying to remember the history of the past while learning to anticipate the future and to set far off goals beyond the existence of a single individual.

Time has a direction and is irreversible to the educated human system.

Like the plants and other animals, young children have no concept of time. New humans have no concept of cause and effect, conservation, density, etc., etc., just as they have no concept of information, matter or energy. The human system is born into this world with an almost empty information storage device. The brain has evolved to learn and it has a lot to learn.

Time, for the human system, has a one directional flow—from conception to death. The ticking of the clock of life is set by the rotation of the Earth on its axis. The tick tock is night and day, night and day, night and day.

Children are conditioned to the tick tock of night and day long before they can read and write. And they can read and write long before they can understand why there is night and day. And understanding why there is night and day is learned by the brain well before the human system becomes a sophisticated time manager.

I have asked hundreds of elementary school students, "Why is there day and night?" From kindergarten through third grade the most typical answers center around the self. All human systems start life self-centered. Most five, six, seven and eight year olds say, "Day is because that's when I go to school, and night is when I sleep." They do not understand cause and effect, so everything is in terms of what they do.

A child's past was yesterday and her future will be tomorrow. Teachers will spend years helping students put together weeks, months, years, decades, centuries, millennium and eons. Students must learn to push history back further and further into the past, and they must learn to anticipate time further and further into the future. All this takes very good teaching and learning, and will not happen without verbal and written communication, and some form of formal education.

Another aspect in the learning about the use of time is the systematic organization, day after day through time, of the school system itself. The very strict, school day regimen forces all children to fit into the metered time slots for a full day of activities. For children who are raised in work and achievement oriented families this school regiment is no problem. These students were reared in a clock watching environment. Entropy reduction comes easier for these students. The management of time at home and at school are not that much different, so they fit in.

Students who were not reared in a clock watching environment often have a much tougher time fitting into the school system. They come from disheveled homes where work and achieving goals are not important. Time management is not stressed and entropy production comes easier for these students. Dishevelment, destruction and disorder are so easy. They take very little time, information, matter or energy. Letting systems go to high entropy fast is very easy—don't do anything.

On the other hand, it takes a long time to put together a forest, a library or a human being, and to keep any one of the systems from going to disorder requires expending and managing the four components of the T.I.M.E. acronym. Keeping systems from going to high entropy fast is very hard.

But, since we desire entropy reduction and organizational skills rather than entropy production and disorderly behavior, teaching about time becomes a part of the entropy ethic. The schools could overtly teach more time management rather than leave it as part of the hidden curriculum.

TEACHING ABOUT INFORMATION

In the T.I.M.E. acronym the "I" represents "information." Every living thing is involved in information management. Without the information in the genes, no organism could grow or go. And for those of us who have evolved senses, we must learn by absorbing information from the outside world.

Once the information is learned and stored inside the brain, we call it knowledge or know-how. Education has taken place. Information is an accountable and we should all learn how valuable it is.

Like in the genes, a photograph for the human eyes, or any other information for the senses:

1. information, genetic or learned, exists in the arrangement of matter.
2. information, genetic or learned, regulates time, organizes matter and controls energy flow.
3. information, genetic or learned, is inversely related to entropy.
4. information, genetic or learned, is created and destroyed.

As we know, energy-matter cannot be created or destroyed, but can only be transformed one into the other and moved about from place to place in the universe.

It would be hard to say that information management is more or less valuable than time management. The acronym must be taken as a whole, because information is just as necessary to build, construct or organize any system.

We may now see more emphasis put on information, because it is being said that we live in the Age of Information. But regardless of the relative importance of this second component of the T.I.M.E. acronym, we must graduate students who are excellent information managers. Information must be very valuable to educators.

If information storage is so intricately bound up in the way matter is arranged, then is the chemistry teacher responsible for exploring this phenomena with oncoming generations? Or does the responsibility fall to the art teacher? Or to the typing teacher, the English teacher, the drafting teacher, or the biology teacher? All of them deal with how matter is arranged, stores and imparts information.

The sad truth is that the textbooks, the curricula, and the teachers spend little, if any, time teaching about information. Oh, a "blueprint" may be mentioned in biology class, as some nebulous set of instructions in the DNA of bean seeds and frogs. The drafting teacher may tell the students to store their "plans" until they are ready to build their house. The English and typing teachers may compliment their students on the "clarity" of their writing. And the chemistry teacher may tell the students that NaCl always "tastes" like salt. Which is like saying that a giraffe always looks like a giraffe, or high C always sounds like high C. All true, but very boring. We should do better for our students.

The kinds of questions we should ask are: "What do genes, brains, languages and computers all have in common?"; "Why has the human system evolved to take such good care of its genes, its brain and its body?"; and "Why should the human system take such good care of its language, its books, its libraries and its computers?"

Though the public schools have not even begun to introduce students to the broad aspects of information science, information management and information values are already a necessary part of our society, every culture, and, indeed, our very existence. DNA management by humans has been evolving for centuries. Humans have worked their will on the plants, domesticated and wild animals, and even themselves by choosing their breeding partners. But now with the mapping of DNA our students face a future of genetic information management hardly dreamed of until this century. Who shall be allowed to live to breed? We cannot and should not avoid these value issues in our society and in our schools.

Education is the management of what information goes into the brain and how. Information stored in the genes and in the brains then becomes the guiding factors for what the brain thinks and for which behaviors the human system will display. The human system creates information when he talks, writes, paints, sculpts, performs, builds, and behaves in any and all manner. And he is taught to do these in a very orderly way. He is not taught to move about randomly, to be a destroyer, to scribble and to make muddled sounds. This is the ethics of education, to reduce entropy. By educating for low entropy, human systems, education strives to form information yielding students.

This ethic of educating to build organizers rather than randomizers helps maintain order in society. Health care, law, government, the arts and the sciences all hold this ethic in common with education, to help maintain order in the society. The problem is that teaching the values of information and order is not something teachers do overtly with their students. The values are "hidden" in the curricula.

As the human system evolves from conception to adulthood there is a striving in the mother's body, in the home, in the elementary school, and in the high school to form each human into a highly organized, complex, low entropy, nonequilibrium, open system with many degrees of freedom.

An uncountable number of things can go wrong, and do. Some students finish high school in a state of confusion, lacking self-control, and bound for disorderly conduct. They have not learned to be good managers of information and high organizers in general. The values asked for in the job market eludes them, and quite often they do not even know why they are unwanted.

Teaching information science becomes another avenue to making more self-aware students. When teachers and students can get their philosophy of life and living in tune with their teaching and learning, then less confusion and more organization will prevail.

I am, of course, advocating teaching students more about themselves as a human system moving away from being randomizers, and going toward being better organized, and better organizers.

My experience with students indicates that when they study themselves in relation to organizing and to randomizing, their minds rise above the usual level of reasoning. Their thoughts *soar* into the philosophy of life and living, into understanding natural phenomena, into understanding ethics and aesthetics, and into a better understanding of themselves.

That is why I call my holistic approach to teaching, and what I have set forth in the book, *S.O.A.R.*, which is the acronym for, *S*tudying *O*rganizing *A*nd *R*andomizing.

I am sure there are many other possible ways to tie the new information science to entropy consideration, to energy efficiency, to matter management, to time management, and to ethics and aesthetics. They will evolve because education has a long way to go to catch up with the world view from the entropy paradigm. Any new curriculum takes much work in research and development.

What I am saying is that my initial efforts have been very rewarding because my students tell me their thinking has been motivated to *soar*.

Information storage, information replication, information transmission, and information retrieval, these are the components of information which allow it to dominate over the second law—the tendency to higher entropy. Information is the overriding factor in nature which makes life possible. Such an important factor must have an evolution, and indeed it does. And the evolution of information in life is repeated over and over again in the evolution of information in each new human system.

First, the information stored in the genes must be replicated, transmitted to start a new human system, retrieved, stored, replicated, transmitted and on and on, one generation after another. The sexual urge, or reproductive drive is built into each one of us as pleasure, a hedonistic instinct necessary for the survival of the information. That must be necessary for plants, pelicans, porpoises and people. If the sex drive goes, that is the end of the species.

This basic necessity for life worked so well (and still does) that the second evolutionary stage of information came into being. The brain evolved as an information storage device, replicating its information by word of mouth and other means of transmission, so other brains could retrieve, store, replicate and on and on, generation after generation. But the brain evolved (and still does) to be more complex than the genes, and more flexible, and more adaptable. Whereas genetic information transmission led to sexual behavior, brain information transmission led to talking behavior, and finally to formal education.

The great survival benefit of the brain can be seen in the rise of humans to rule the planet. Making maximum use of the brain's capabilities is indeed power. And those individuals, families or nations who are educated to the fullest rule the ignorant who have ignored their brain's capabilities.

The evolution of the genes to the brains has led to a third stage of information evolution—the human system organizing matter *outside* the system: the painting, the printed word, the musical score, the film, the tape, the record, the instruments, the mass produced objects, and finally culminating in the computer, a raw storer, replicator, transmitter and retriever of information, with the basic attributes of the genes and the brains, yet not bound up in any transient body.

The computer's evolution depends on the educational development of the brain. From the genes, to the brains, to the arts and sciences, to the computer, the evolution of information storage, replication, transmission and retrieval marches on. But for the evolution to continue, education becomes an added necessity, because the later stages of any evolution cannot exist without the former. The genes get on quite well in plants without brains, but the brains are dependent on the genes. Likewise, the brains get on quite well in animals without computers, but the computers are dependent on the brains. Or, said another way, the information attributes of a computer would not exist without the information attributes of the genes and the brains. That makes some kind of sex and some kind of nutrition and some kind of education necessary if information is to continue to evolve.

Every student in the United States is exposed to the two laws of thermodynamics. Clausius set down the very terse versions of the energy laws in the 1850s.

1. The energy of the universe is constant.
2. The entropy of the universe tends to a maximum (Kubat 1975).

Today, with more accumulated knowledge, teachers and textbooks use broader definitions:

1. Energy-matter cannot be created or destroyed, but can only be transformed one into the other and moved about from place to place in the universe.
2. Energy-matter will tend to maximum randomness in any isolated or closed system.

There are, of course, many other ways these two laws can be, and are, expressed. But, whatever the expressions, the basic concepts remain the same:

1. Humans have found no way to get energy-matter from nothingness, and no way to annihilate it.
2. All systems tend to go to disorder eventually, even the most complicated, living, open systems which maintain and repair themselves by taking in and expelling energy and matter as long as they live.

What gives living systems, especially humans, this measure of success in spite of the entropy law? That is what information theorists ask themselves. And that is why information theory is coming to play such an important role in understanding life's place in natural phenomena.

What has been discovered are a new set of laws which necessarily evolved for life to cope with the laws of thermodynamics. Analyzing both sets of laws helps students better understand life's evolution.

The first law of information: INFORMATION, GENETIC OR LEARNED, EXISTS IN THE ARRANGEMENT OF MATTER.

The arrangement of matter yields us information and determines our form and function, as it does for any other living system. What makes humans so different from other life forms is their ability to arrange matter outside themselves, and to recognize it as representing something which it is not. Pictures, films, records, photos, models, graphs, charts, maps, words, numbers, etc. are an infinite variety of symbols to represent the rest of nature. A painting of three trees is not three trees. The words "three trees" are not three trees. And the number "3 trees" is not three trees. Rather, they are all ways for us to communicate with each other about trees. They are ways of symbolizing and classifying nature, so that nature becomes manageable to the human brain.

Since there are an infinite number of forms and functions, properties and relationships in nature, humans assemble into groups and categories those representations which will be the most beneficial to each oncoming generation. We start with each child's immediate self and environment. This is what each language, each culture and each family does for their new generations. It is necessary for survival that, with each language, nature becomes classified and manageable for communication and for the higher education of each new human.

Education is responsible for human beings acquiring higher and more complex communication skills. But the infinite number of symbols cannot simply be dumped randomly into the brains of children. Education also has the task of organizing what goes into the brain for the better building of communication skills. This is the difference between instilling confusion and understanding. Educators must help each student understand the symbols which stand for other natural phenomena. This is how the tree of knowledge, awareness and logical reasoning grow best inside the human brain. And it is because the roots of knowledge are in the arrangement of matter, the residence of information.

The Second Law of Information: INFORMATION, GENETIC OR LEARNED, REGULATES TIME, ORGANIZES MATTER, AND CONTROLS ENERGY FLOW.

If energy is to be stored rather than scattered; if energy flow is to be slowed down or speeded up; or if energy is to have its direction of flow channeled, then the organization of matter by humans is imperative.

The plants discovered this simple fact long before any humans were around to become aware of it. Oh, that the human system could trap the scattering energy from the sun and store it so easily! However, like the other animals, we cannot eat direct solar energy. For our own bodily growing and going, we are completely dependent on the entropy decrease taking place inside the plants.

Like the work of the chlorophyll in the plants, the information in the genes of all living systems builds the systems (arranges the matter) in order to control energy storage and flow. There is no other way to existence for life. Fortunately, we do not have to think about building our circulatory systems, or any other of our subsystems which make us whole. What a task that would be!

As humans become more energy wise, we are seeing the necessity for education to place more and more knowledge about energy into the curriculum. Humans have now also come to realize that the information entering the brain (via energy transmission) has enormous control over how each person uses his energy in behaving. In fact, any considerations about human behavior are considerations about how the human system uses energy.

The Third Law of Information: INFORMATION, GENETIC OR LEARNED, IS INVERSELY RELATED TO ENTROPY.

If there is not some structure to matter, some organizational arrangement in matter, then there can be no energy control. In other words, without the organization of matter (which is the same as high information and low entropy) the energy laws tell us that energy rapidly disperses to an unusable state. Energy scatters from an area of high concentration to one of low concentration. We see this one directional flow from a burning candle, from a sound source, from any mechanical device, from an electrical system, and fortunately for us, from the sun. If that tendency to flow toward equilibrium is to be reversed, then additional time, energy and matter must be used from outside the system. In addition, know-how, which comes from information and reasoning, must be applied.

Human engineering performs this feat quite well, as can be observed in the example of the refrigerator-freezer. Just plug in the electric cord of the refrigerator and the heat energy inside is moved to the outside. Inside, the entropy will decrease. A tray of cold or frozen water molecules has more order, less entropy, than the same hot water with molecules rapidly evaporating. But outside the refrigerator, all the energy leaves as heat, and some unwanted sound, going out into the space around the earth. The energy is lost from further human usage, except for warming the kitchen a bit along the way.

The point is that humans have learned how to reduce entropy locally. The know-how comes from the human brain. We see that even a simple closed system like a heat pump can have its internal entropy reduced. But, as we are all aware, the refrigerator, like any other functioning device built by humans, will eventually wear out and deteriorate.

This is precisely what all living systems have had to contend with. They have had to control the flow of energy to exist, because they have had to reduce entropy to exist. But unlike the education needed to develop the engineer, the genes evolved the same know-how billions of years ago, with an added dimension not yet simulated by human engineering. Since any individual functioning system will eventually wear out, something more is needed for prolonged survival. The genetic solution was, and still is, to assimilate usable elements into living systems, along with energy. This process has led to an infinite variety of open systems. Therefore, unlike the closed systems built by engineers, the living, open systems are able to make internal repairs.

But there is still another aspect of the living, open system which is just as important as internal maintenance, and perhaps even more so. The information in the genes must be reproduced for each cell in a living organism, and it must also be passed along from one generation to the next or else the whole process of reducing entropy would cease. This leads to the possibilities for constantly changing evolution built into each one of us, into each species and, indeed, into all of life on earth.

Because the organizational arrangement of matter (information) is the key to prolonged survival, it is not surprising that the brain has evolved to such prominence in living systems. But the brain evolved enormous flexibility, unlike the fixed genetic information each one of us inherits at conception. Throughout nature there are infinite possibilities for the arrangement of matter. If the human brain were to cope with all the possibilities, the best survival tactic would be to have as few restrictions as possible on what information the brain could accept.

We see the flexibility of the brain when we conceptually survey all the past and present human cultures. Think of the number of different languages. Because the brain is so flexible, an infinite number of written and spoken words are possible. They represent the infinite variety of forms and functions in the rest of nature. This accounts for the many different types of beliefs, customs and traditions expressed in human cultures. The human brain, like the genes, is geared for adaptation and change . . . but can work at it more rapidly than the genes . . . a tested way to survival.

The genes have been wending their way through the branches of biological evolution for billions of years. The genes found a way to utilize energy and matter for prolonged survival. Now the human brain is becoming aware that it too must work with the same natural phenomena, if the human system is to be successful in continuing to inhabit this planet.

The Fourth Law of Information: INFORMATION, GENETIC OR LEARNED, IS CREATED AND DESTROYED.

When a library is burned down, or when a tapestry is eaten up by moths, or when a child's quilt is dissolved in acid, the energy and matter of which the items are made are not destroyed. There is no way to create or destroy energy-matter. But the buildings, the books, the symbols, the pictures, the numbers and the child's quilt can be destroyed. They can be sent to high entropy fast, and that is not often to our liking.

The organized matter is what makes the world around us manageable and meaningful. When this organized matter and concentrated energy (information and locally reduced entropy) are lost from our further usage, we do not weep because the energy and matter have been dissipated and dispersed. Of the billions of people who have existed, most of them have not understood that energy-matter cannot be created or destroyed. But still, humans have had to make organized matter and organizing matter very valuable to them. It should remain so, and is one of the most basic values educators can pass along to their students. Why?

1. Because of evolution and communications, as we have seen.
2. Because the cost of replacing organized matter can be very great in terms of more money, time, knowledge, matter and energy required.
3. Because the immediate loss of organized matter can threaten the very existence of the loser.

Deterioration, decay, destruction and death all remind us how fragile existence is. Better to punish the destroyer and praise the creator!

TEACHING ABOUT MATTER

In the T.I.M.E. acronym the "M" represents "matter." It is matter which must be arranged to hold information, in the genes or for the senses. Randomly arranged matter is not very valuable to us genetically or educationally. This leads to disorders, and disorders are not conditions we cherish.

There is a place for everything and everything should be in its place. Teachers must demand this rule be followed just as must the librarian, the bank president, or the captain of a ship. There will not be efficient time and information management without good material management.

This third component of the T.I.M.E. acronym is just as important as the other three. I cannot imagine a teacher training students to distribute books around the library randomly. Bankers who cannot keep their money and files in a state of low entropy will not survive the banking business. And the ship's captain who insists that all material objects on his ship exist in disarray and chaos will rapidly be replaced.

Organizing matter is important. Its importance for educators extends from putting the letters down on this page to an orderly system of arranging books and all other supplies in each classroom.

Organizing and learning to organize matter is probably just as important as the content of a course. (Which needs to be organized too, before it is presented to students.) This is so because regardless of what occupation a student may choose, his employer would like for him to have learned to be a good organizer. A student's record which would classify him as an excellent organizer would certainly be helpful to a potential employer.

The periodic table of the elements is the place to begin an understanding of nature's "building blocks," including the human system's composition. Each student is entitled to an understanding of the elementary particles, the atoms, and the molecules they form, leading to the information thus made available to the human senses. Only through this basic foundation of matter's microscopic composition can an appreciation of the macroscopic world be in tune with reality.

Matter is necessary for the existence of everything. Matter accounts for our planet, for the states of matter, and our building materials; our breathing; our atmosphere, hydrosphere and lithosphere; our plants, animals, and the information available to their senses; our science, technology, and all the arts. How could we develop humans who do *not* love matter, respect it, and want to study it, understand it, and appreciate it?

A disrespect for existence seems to me to be a terrible flaw. Something must have gone wrong in our educating students about matter. It is appalling that so many humans could reach adulthood in the United States without an understanding and appreciation of matter. Where do we go wrong?

Every high school in the United States offers at least a year of chemistry. Why are the teenagers not thronging to the course to be enlightened about existence? Where is their reverence for what they are made of, and the composition of all else about them. Something must be wrong!

Something is wrong! We teach chemistry in our schools as though we were trying to turn each student into a chemist. The scientific method is the primary concern. Over and over the students are badgered again and again about using the "correct" scientific method. They must practice the scientific method through elementary, junior and high school, and on into college. They get tired of it, bored with it, and reject it.

No experiment is valued more than another. It is only the "correct" experimenting which counts. For all of their early years, the content of chemistry, indeed all of science, is thrown to students randomly. Few students acquire a logical understanding of nature and how it works. Then how can we expect a respect for the matter which accounts for our existence? Few students gain an appreciation for what science has discovered about natural phenomena.

Most students walk away from school with random bits of facts, laws and theories about nature, tucked away in the disorganized information storage centers of their brains. Do we think students will somehow, miraculously put these randomly acquired bits and pieces together after they reach adulthood? Experience tells us this does not happen. Given this random treatment, most adults will stay confused about nature, science, and indeed, about the value of their education. And we will have underdeveloped and confused minds walking the streets on welfare while our nation needs logically thinking people whose minds are organized in an understanding of nature and how it works.

How did we go wrong? In our schools, science courses got specialized into physics, chemistry, geology, biology, and a few others, such as health. The students get to grapple with a few pieces about the earth somewhere, about energy here, about matter there, about life here and there, and about diseases everywhere.

By the time they get to high school, teenagers' minds are capable of soaring. They need abstract thinking, big concepts, the picture of the universe and how it works. We take them into chemistry lab, offer them a cookbook formula for extracting hydrogen from water, force them to memorize and balance some equations, tell them they have learned how to *do* science, and bore the hell out of them because they have been through it so many times before.

They do not want to do it again, and probably will never need to. They do not want to get into specific thinking again, when their brains are hungering for holistic thinking. Oh, sure, the best and brightest high achievers will take the course to get into college, medical school, engineering and the like. But most students play the avoidance game. As they say, "Who needs the anguish?"

We need to develop holistic courses for the majority of high school students. A holistic course about nature lets their minds soar. They like it. They learn. And they tell you so. During holistic teaching, students claim to put their philosophies of life, living, and learning into an orderly perspective. Current chemistry courses and textbooks come nowhere near such a rewarding goal.

Finally, when order versus disorder is emphasized in a holistic course, a foundation for ethics is laid down. From it the students learn to appreciate why humans have loved and rewarded the creator, not the destroyer. Ethics, aesthetics, education, religion, and science make sense to them. And they tell you so!

Those are the rewards for the students and the teacher when guiding adolescents to holistic thinking about nature.

It would be silly for me to try to name everything which is made of matter, for I would have to name everything.

However, there are many important concepts regarding matter which need to be studied in a holistic course. I have included some of them in this book: matter's relation to entropy, organizing versus randomizing, order versus disorder in the universe, energy storage and control, information storage and control in the genes and for the senses, and systems classification and analysis.

There are, of course, many more concepts such as a student's ability to read and understand the periodic table of the elements, which will go unanalyzed. There are also ideas about matter from the history of human development, which would be worth pursuing. One example which students like is the concept of the distribution of matter on earth.

When students look at their environment today, they see the glass, the brass, the books, the bricks; and the silicon, the copper, the aluminum, the gold, the silver, the mercury, et cetera. It is good for them to contemplate how humans came to accumulate enough of the elements in one place so they and their children's children could have the objects of matter which make our existence easier and more worthwhile.

It would have been nice if humans could have evolved to a state of self-awareness on this planet, and at the same time found all the elements neatly stacked in one place ready for human ingenuity to creatively use. Here all the atoms of copper correctly labeled with instructions for all its possible uses. Next to it all the mercury atoms with its warnings and uses, and on and on.

But no! How did it really come to pass, this affluent planet with all of its transport systems of wires, pipes, cans, cars, pockets and plans? Scattered randomly all over our planet, the elements have had to be ferreted out and accumulated and organized by the hard work of humans over thousands of years. That is part of the reality of reducing entropy on this planet. Much time had to be spent, much energy had to be used and sent irretrievably off into the universe increasing its overall entropy, and know-how? Humans could never have done it without the evolution of the brain, intelligence and education.

That is progress. That is locally reducing entropy. That is organizing matter on a massive scale. And it still goes on. It must go on. We value matter, and we value organizing it. That is much of what human behavior is all about.

Best we teach it the way it is. The ethics will follow therefrom.

The understanding of matter, the accumulation of matter and the using of matter to make more and more complex closed and open systems with energy flow through (next chapter) fascinate the students' brains. Their thinking processes *soar*. Teaching about matter from a systems point of view is not boring.

Systems within systems within systems. No doubt part of the reason why so few students study systems in any organized way is because so few teachers know how to teach systems in an organized manner. A general teaching rule is that students learn by developing from the simple to the complex. This is true for our biological, our behavioral, and our brain development. And this is true for the evolving of our species, as well as our individual development.

Another general teaching rule is that students learn by developing behavior and ideas from what was past, to what is now, to what might be. Following these two guidelines, I should not get into too much trouble by starting with our Milky Way Galaxy and our Solar System and moving through time and evolution to the human central nervous system. Both rules will thus be fulfilled, the teaching through time and evolution, and the teaching from simple to complex. My experience with high school and college students reinforces this order of teaching systems.

Remember, there are three basic kinds of systems. There are isolated, closed and open systems. Each of the latter evolving from and inside the former.

The universe is probably an isolated system. That is it exchanges no energy-matter with anything outside itself. If this is true, then there exists a fixed amount of energy-matter existing in a boundless, infinite space-time.

The Milky Way Galaxy probably evolved from the big bang, just as all islands of energy-matter have done in the universe. While the big bang was a scattering phenomena, our galaxy became an organized system held together by gravity. Some long lasting order began to display itself as the galaxy's stars came into existence. The stars emit enormous amounts of energy from inside the galactic system, and by contrast the galaxy absorbs only a minute amount of energy from outside itself. What this may mean is that all the energy from all the stars in all the galaxies eventually will become equally distributed throughout the universe. The universe will then have reached its maximum entropy, its maximum equilibrium, and no more growing and going will be possible from such a state. Maybe? In this holistic sense, the universe is a rather simple, isolated system, heading for a heat death, going toward maximum randomness, from which no return to order is possible. Maybe? Stay tuned in to the cosmological findings.

In the meantime, however, before those billions of years pass away, for the here and now, there are plenty of opportunities for orderly, closed and open systems to emerge.

Our solar system is one such emerged and evolving closed system inside our galaxy. Like our Milky Way, our solar system emits enormous amounts of energy into the universe. By comparison, very little energy comes into our solar system from the rest of the universe. We must live with what energy and matter we have available here and now.

Within our solar system, our home planet, Earth, has emerged and evolved as another closed system. The amount of matter making up the earth is relatively fixed. (Except for a few entering elementary particles and atoms, and some occasional bits and pieces of meteorites.) But the energy comes into the Earth's closed system from the sun and leaves the system in balanced proportions. Much like the energy striking and leaving a mirror. This balance between energy input and energy output was, and still is one of the necessities for the emergence and evolution of life on earth.

Within the closed system of Earth, (energy moving in and out) with a fixed amount of matter available, an infinite number of new systems could emerge and evolve.

Life became different from the rest of constantly changing matter by developing from and into more and more complex, open systems. In such systems, (exchanging energy and matter with the environment) complexity would reach new heights.

To accommodate the need for energy and matter input and output, systems arose inside living systems. In animals, complexity increased further with the advent of the digestive, the circulatory, and other internal systems. The highest, and the most complex evolved system, inside the animal, open system, would become the central nervous system.

The rise of the central nervous system in open, animal systems created the demand for learning in lower animals, and finally accounts for why formal education came into existence in the human species. The human system, with its central nervous system, and more organized learning, then became capable of building more and more complex closed and open systems. They range from simple machines and musical instruments, to printing presses and nuclear ships, to batteries, hotels and hospitals; from clans and bands, to banks and governments, to computers and robots, to orchestras and cities, to school systems and school buildings and classrooms.

The human system is responsible for the evolution of more and more complex closed and open systems. And the human systems are responsible for the management of the systems they have created. This responsibility cannot be fulfilled without an education in systems management. An area we cannot long neglect in the public schools.

TEACHING ABOUT ENERGY

In the T.I.M.E. acronym the "E" represents "energy." We are constantly being bombarded with information encouraging us to conserve energy, or not waste energy. Using only the required, right amount of energy to do a task as efficiently as possible is something we cherish. This value too is no less important for managers of classrooms and schools, than it is for ships' captains, bank presidents, engineers and architects.

When we organize time, information and matter, we are also assuring ourselves that energy will be properly managed. The four components of the T.I.M.E. acronym are inextricably interrelated. Were we to go about our teaching tasks in a randomizing, disorderly manner, we would certainly be wasting our energy and the energy of the students placed in our trust.

When time is left unplanned, and information is randomly shoved into the developing brains of oncoming generations, and materials are kept in constant disarray, then we treat our student's brains and behaviors to confusion and chaos. Energy is then wasted, because it is logical thinking and coordination which build the efficient, highly organized and highly organizing human being.

This concept can be applied to any discipline and to discipline itself. A very large part of the hidden curriculum is for educators to bring order out of chaos. We strive to develop students who are entropy reducers, not entropy producers. Or, said differently, a society will not survive for long, the same as one individual person will not survive for long, in a state of confusion and chaos, with destruction and disorder as a value while existing.

An example I use with my future teachers is the randomly assigned teacher who was told to build a band on Saturday morning. Saturday afternoon he randomly passed out one hundred musical instruments to one hundred randomly chosen students. Saturday night the new band director and the new band did whatever they could, or whatever they wanted to do half way through the football game. The school board and the community were not impressed. What society will accept, and what we call ethically and aesthetically good, all can be merged under the T.I.M.E. acronym, and the entropy ethic.

How should we go about the business of teaching another person about energy? That is one of the central concerns I must address here. A logical, well-thought-out plan is needed. I hope the approach I offer here can fulfill that obligation.

Since I am concerned with the human system, and each student feeling and knowing he is a part of nature, my approach centers around the human system. I call it the student-centered approach, because it is human system-centered. Students generate innumerable, creative, energy flow drawings going back to the sources of their own energy.

There are also innumerable energy flow diagrams which are created to represent other happenings in nature. But I begin by putting each student squarely in the middle of natural phenomena. This helps the students understand that each of us is dependent on sources for energy, energy transformations, and the forms energy assumes as it moves from place to place.

Teaching about the sources of energy, the forms of energy, the ways to store energy and the transformations energy goes through, should precede the study of the energy laws, the study of systems, and the study of the human control of energy. Though I have found that in the actual classroom setting, the quick students' minds rapidly flit from basic energy considerations to ethical considerations. A sharp student's mind moves ahead very fast, in any subject, when you move from the simple to the complex in a very organized and logical manner.

The point is that each human should learn to follow the flow of energy from its source to the point where it is irretrievably lost, including all of the transformations between.

This task can be accomplished with the help of diagrams before adolescence. Holistic concepts of this kind can be best learned with diagrams. A picture is worth a thousand words. Or, said another way, you could talk all day to a novice elementary student about the energy getting from the sun to the student's behavior. When finished, the words alone might leave many students in a state of confusion. But by drawing the picture first, the student has a logical, accurate representation in which to fit later pieces of knowledge.

An analogous example is when children (or any of us for that matter) are asked to put complex puzzles together, with and without a peek at what the whole, finished picture will look like. The puzzle goes together faster, and more accurately, when the picture is kept in view while fitting the pieces together. So, I have students draw a picture they can use as they go through the process of learning about energy flow.

Later, as the bigger energy words, the energy facts, the energy experiments, the energy laws, the energy systems, the energy control, and eventually for some students the mathematics of energy unfold, they fit neatly into the holistic concept the students have retained, in their brains and in their paper diagrams. Of course, this process is not new to education, but rather a principle too few teachers have applied well enough in teaching about energy.

My experience has taught me that any normal teenager can move from entropy to ethical considerations related to energy. My experience has also taught me that not every student is going to tackle the mathematics involved in all of the energy transformations. Thus, I find just cause in taking each student as far as he can go in understanding energy, with or without mathematics.

Lindsay's first sentence in his introduction to *The Control of Energy* is, "Energy is the premier concept of science and probably the most important single idea the human race has developed in its attempt to describe and understand experience."

Yet many educators, myself included, and scientific associations, find the state of energy education appalling! It is time we begin to do much about it.

Here at Texas State University, in the last half of the 1980s decade, we embarked on a new effort in energy education. Working with the Seguin Independent School District under a cooperative grant titled Curriculum Development in Energy Education Project (C.D.E.E.P.), I began teaching all Seguin fifth and sixth grade science teachers to classify, on a color coded and shape coded arrows, the sources of energy, the forms of energy and the ways to store energy. The teachers then had all their students cut out their own arrows for classifying energy and proceeded with many activities to teach the students energy classification and how to conceptualize energy flow.

The project was very successful. Four more grants were funded to continue the project. Summer classes were held in our physics department to train more science teachers from other school districts. All seventh and eighth grade science teachers in Seguin were taught the process. An artist was funded to make renderings of energy flow for overhead use in classrooms. Science teachers from private schools and more school districts arrived in more summer physics classes to learn the Seguin Model (as it was called in education circles) of teaching energy flow conceptualization. More teachers were trained in San Antonio to train more teachers of science south to the Mexican border and west to Del Rio.

What accounts for such success in teaching about energy? Elementary school science textbooks had the sources and forms of energy in them for teachers and students to read about. But as hundreds of our university students and hundreds of middle school students and their teachers, grades four through nine, were tested, no one could classify the sources and forms of energy. Obviously, teaching about energy has not been important to these science teachers and their students. Then these students move on to high school and college unprepared to study more about natural phenomena. Physics is avoided and the other physical sciences too. And the way is lost into the energy and systems professions, where so many people are needed. What is even more crucial, as I have stated several times in this book, we in the United States are taking our students through schools to adulthood without a firm foundation in ethics.

Our energy flow project is so successful because the teachers and their students like it, they understand it and they want it. It opens a whole new way of thinking when they learn to conceptualize energy flow.

When a proud sixth grader came to class one day and said she was going to run her rat off gravity, I knew something was going right. The student went to the chalkboard and drew a running rat and wrote kinetic energy under it. Working from right to left, as she was taught to do, she drew an arrow going into the kinetic energy from muscular energy. Then she drew a head of lettuce and said it had energy stored in its chemicals. She put in an arrow from the lettuce to the rat. Then she drew a light bulb and put an arrow from the bulb to the lettuce, writing light energy above it. She also said that was where photosynthesis took place and the lettuce was growing in a greenhouse. She labeled the arrow going into the light bulb electrical energy and its tail ended at a little generator and turbine she drew. She said that was mechanical energy and they were turned by the moving tidal water. After putting kinetic energy prior to the turning of the turbine, she placed a big, printed GRAVITY at the far left of her arrowed diagram. She said she could run her rat off the energy from the sun, or the energy from nuclear fission, or from the heat energy from the earth, or from the wind turning a windmill. I knew she could. And I know she was far ahead of the science teachers, who are graduate students in my classes each summer.

All but the slowest of students learned to conceptualize energy flow. Once they got the hang of it, they wanted to do every energy dissipating device they could think of. They asked to do bicycles, cars and electric golf carts going down fairways. They did a piano being played, brought in wind-up toys, music boxes and all kinds of appliances. And the students wanted to spend more time on the project than their teacher could allow.

By the summer of 1989, we prepared two more physics classes of middle school science teachers to be able to teach energy flow conceptualization. Hundreds of students in Seguin have been through the procedure of classifying energy and conceptualizing energy flow with different teachers. They still love it and so do the teachers.

A feeling for and a knowledge about energy are not mental states acquired independently by each new human system! Humans have been on this planet a long time, but it took the rise of science to find out about *The Control of Energy*. Literacy about energy can only come about through a well planned educational process. So, we must begin and build and correct that process–which is what is happening now in the United States.

The knowledge of energy and its relationship to entropy and ethics will not go away. But when the human system tries to operate without that knowledge, he is surely operating in the shadows and not as an enlightened one.

"The relationship of energy to entropy is so glowingly obvious now, I don't know how I missed it all these years," a retired military officer (graduate student) told me at the end of a semester. "Now I am just beginning to understand what the striving for energy conservation, efficiency, and time management was all about." He went on to say he had thought it all had to do with cutting cost or wisely managing money. But now he sees life more fully and more deeply; he is hooked on the entropy concept, and he has more to offer his students then energy management based on the money god.

The diagram on the next page (from one of the 1988 booklets) was conceived in 1988 by Helen Lawrence with her fifth grade science class in Seguin and artistically refined by our university press. The covers of both 1988 booklets, selected by the other teachers in those classes, are published in the appendix of this book.

INFORMATION AND EDUCATION

The human system must learn. Human biological development and learning are inseparable. Education is the art and science of helping the human system learn. The curriculum is what the human will learn. The nation, the state, the parents and the school administrators hire teachers to facilitate what, when, how and where the developing human must and will be taught. Teaching the human system is the job of the professional teacher in the provided time and school.

Teachers are trained to be certified professionals. They are educated for their tasks in accredited four year colleges and universities throughout the nation's states. Helping developing humans learn has become a very necessary, and should be a very noble, profession.

Yearly, new and old professional educators come together with students in the schools. The teachers know who to teach, how to teach, where to teach,

A CURRICULUM ACTIVITIES GUIDE

FOR IMPROVING INSTRUCTION IN 5TH AND 6TH GRADE PHYSICS

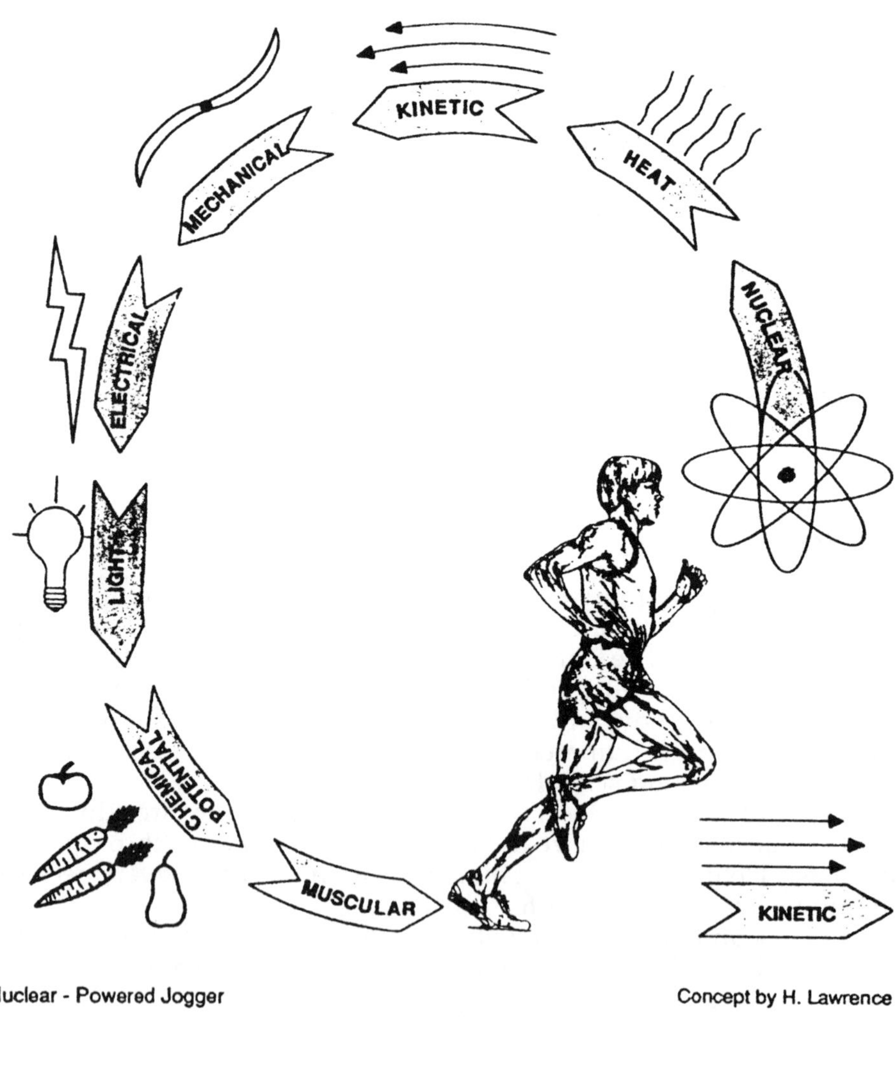

Nuclear - Powered Jogger

Concept by H. Lawrence

why to teach, and what to teach. A few teachers are excellent, some are very good, most are average, some are not so good, and a few teachers are bad. This is not so different from ranking people in any other lines of work. What makes teachers so different is that they are in the business of directly influencing the brains and behaviors of developing human beings. Of course, many other members of society also help mold the thinking and conduct of each human system, but it is the certified teacher who is the trained professional. The teachers are trained to know the best ways to mold the brains and behaviors of oncoming generations.

This is precisely why every special interest group wants teachers to teach to a special way of thinking and behaving. Individuals and groups with "axes to grind" know they must "get to" teachers through the boards of education, administrators, legislators, PTAs or some other way, if their special beliefs, desires and ways are to be instilled in new students.

By the very nature of the profession, teachers are very deeply involved in building good and bad thinking and behaviors. Teachers must decide what is right and what is wrong as they influence the brains and behaviors of each developing human system. Teachers are very deeply involved in teaching ethics. Teachers are the professional inoculators of what their society deems right and wrong in thought and actions. The student is influenced to fit the wants and needs of the society.

Mary Poppins—Oh! That we could be as good at teaching as she! We can come close by knowing our subjects well, by being consistently fair, and by having genuine care for each student. But there is much more.

Our students deserve stimulating, challenging and obtainable goals. A good educator shares those goals with her students, and together they march along thirsting to drink from the well of knowledge. But there is much more.

The good educator guides, encourages and evaluates. The good educator's students know daily, weekly and monthly that they are marching down the road to success. Students' exercises, papers and tests are promptly returned to them. The good educator replaces trickery and deceit with wit and wisdom. But there is much more.

Students know a good educator wants them to succeed. Students know they are loved and they are important in the evolution of humans and the evolution of nature.

What students are *not* being taught is much of what this book is all about. There is order and disorder in nature. There are entropy producing and entropy reducing activities going on all about us all the time. The human system is a part of nature and, therefore, a part of the ordering and disordering processes which are omnipresent.

Yet, in our schools, because of the way the curriculum is arranged, there is very little, if any, attempt made to educate oncoming generations to ordering and disordering concepts. We live and rear our young in an ethical education void.

Of course, the concepts are there in astronomy, geology, physics, chemistry, biology, psychology, sociology, business, music, physical education, art, etc. But this dividing of nature into bits and pieces means the order and disorder concepts are taught to students randomly—hit or miss—and rarely as a coherent, logical, important consideration about life and living. So, while most of life's occupations center around striving to keep entropy from increasing inside systems, the human system's brain will not easily get to the entropy ethic under such a procedure. What I recommend and delineate here is a new, required course and textbook called, *ORDER AND DISORDER IN NATURE,* with the students and the entropy ethic implanted in nature and in the curriculum.

Time management, information management, matter management, energy management and systems management. When we learn to manage these, we are learning to manage entropy.

This is the wave of the future in education. Sadly, our secondary school systems' classrooms, content, textbooks, and teachers are far from producing systems managers. Much of the curriculum may be even going in the wrong direction. This I should expound on. For it may explain too, why so many students are bored and ill prepared for life after school, as well as not having a firm foundation for ethics.

The teenage years are the time when the human system's brain is capable of soaring. (*S*tudying *O*rganizing *A*nd *R*andomizing *I*n *N*ature would be another good title for such a course.) The adolescent can reason abstractly. He can think holistically, taking many possibilities into account. Yet, much of what he is expected to learn in the schools goes in the opposite direction. Instead of studying a system holistically to see how it fits in with a philosophy of life and living, the students are forced to dwell on the more and more minute parts of a system. Students are burdened with *memorizing* the names of smaller and smaller parts and the names of their functions.

Let us look at the human system to see how this process of teaching works. The lesson begins: The human and all other living things (except viruses) are composed of cells. This is the general statement, which starts with the macroscopic and moves immediately toward the microscopic.

The lesson continues: There are procaryotic and eucaryotic cells. Inside these cells there is protoplasm and organellas. Inside the protoplasm there is the nucleus and cytoplasm. The cytoplasm is made of a matrix, plasma membrane, endoplasmotic reticulum, ribosomes, Golgi apparatus, lysosomes, etc. Soon all students are immersed in the memorizing of the names and functions of more and more specific parts.

Since the science textbooks, curricula and tests are built to cause memorizing more and more specifics for grades kindergarten through twelve, the science teacher has no time to teach for a comprehensive understanding of natural phenomena with a firm foundation for ethics as a part of the goal.

Yet, as I have tried to make clear in this book, scientists from Einstein to Prigogine have shown us how and why we should embark on a new avenue in education in general and science education in particular.

I have tried to delineate what that avenue is in this book. Throughout the histories of hundreds of races of the human system in isolated civilizations it is shown that the human system is capable of learning anything in any language. Also, the human system's brain is geared for adaptation and change. Now it is time, once again, to change the way we are educating the human system.

PART 5

Appendices and Bibliographies

APPENDIX A: SOME STEPS TO UNDERSTANDING THE THERMODYNAMIC IMPERATIVE

Defined:
1. Thermodynamics is that branch of physics which deals with the relations between heat and other forms of energy, and with conversions of one form into another.
2. Imperative means "not to be avoided."

THE FIRST LAW:

Energy-matter can neither be created nor destroyed. (It is necessary to be able to follow the flow of energy from its source, through its several possible conversions, to its final dissipation into the space around the earth.)

THE SECOND LAW:

Energy-matter has a direction of flow. (Energy flows from an area of greater concentration to one of less, *unless additional energy is supplied from an outside source to overcome the normal flow.* A refrigerator is an example of human manipulation of energy and matter to defy the usual direction of flow.)

(As matter moves (kinetic energy) in its direction of flow, it can never be totally converted from one form to another. Always some energy will be lost to the surroundings in the form of heat. Heat loss or energy dissipation is the unavoidable result of friction.)

(Thus, the second law explains why a perpetual motion machine is not possible.)

(Another necessary concept revealed in the second law is the fact that all energy-matter tends toward maximum randomness. Therefore, the law is sometimes called THE LAW OF INCREASING DISORDER, and contains within it the concept of entropy.)

ENTROPY IS THE MEASURE OF THE AMOUNT OF DISORDER IN A SYSTEM

$$S = \text{entropy}.$$

"Entropy is related to disorder:

$$S = R/N_A \ln\Omega,$$

where R is the gas constant: PV/nT; and N_A is Avogadro's number: 6.02×10^{23}. Omega (Ω) is the number of ways a thing can be mixed up inside, microscopically, and still look the same outside, macroscopically. It is a measure of our lack of information regarding a system's submicroscopic condition. When two liquids mix, or a moving mass produces on impact that random thermal motion we call heat—when, indeed, any natural event occurs—then Ω (and S) increases.

"Systems and their surroundings, jointly, move spontaneously from low—Ω states to higher Ω states. *Entropy is more apt to be got than lost and is always on the increase.*

"Whenever the range of micro-states accessible to a system is diminished (by, for example, unmixing a mixture of water and sewage, compressing air into a tire, or cooling a diamond in a dewar) an increase is achieved in information regarding the system's internal condition, Ω is diminished, and, by Relation (2), the system's entropy or "disorder" is diminished. Elsewhere entropy must increase."*

(Any orderly system will move toward disorder unless additional energy is supplied from an outside source to overcome the normal flow. Life on earth is an example of genetic information manipulating energy and matter to defy the usual direction of flow.)

LIFE ON EARTH REDUCES ENTROPY.

From the simplest, replicating virus, to the most complex civilization devised by humans, there is a continuous striving to ward off entropic doom or disorder. If this is to be accomplished, or if order and life are to be maintained, then the laws of thermodynamics tell us there must be additional energy constantly supplied from an outside source to overcome the normal flow. Now, almost all of that energy comes from the sun.

Before continuing, it is necessary to point out some pertinent concepts from the evolution of life on earth:

All living systems are a part of nature.
All animals are living systems.
All mammals are animals.
All people are mammals.
Each person is a part of nature.

Life on earth has evolved from a state of disorder. And each life still does evolve from more randomly moving atoms and molecules. Trapping the energy from the sun, complex chemical bonding, or orderly arrangements of atoms to highly organized molecules was possible, and still is.

---

* Henry A. Bent, "Entropy & the Energy Crisis," *The Science Teacher* (May, 1977), pp. 29-30. Besides the formula relating entropy to disorder, Professor Bent's article contains many of the energy transformation equations.

An orderly arrangement of matter, efficient use of energy, replacement of worn out parts and individuals, and an outside energy source, are necessary ingredients for a living system.

Nothing moves without using energy. Individual, living, human beings move. Individual, living, human beings use energy.

The above is merely the basis for the thermodynamic imperative. Or, it is an abbreviated background for saying that in any considerations about human conduct or behavior, the laws of thermodynamics are "not to be avoided."

APPENDIX B: MAXWELL'S DEMON

There is more disorder, more random distribution, more equilibrium, and an equal temperature throughout both chambers when Maxwell's demon is not operating.

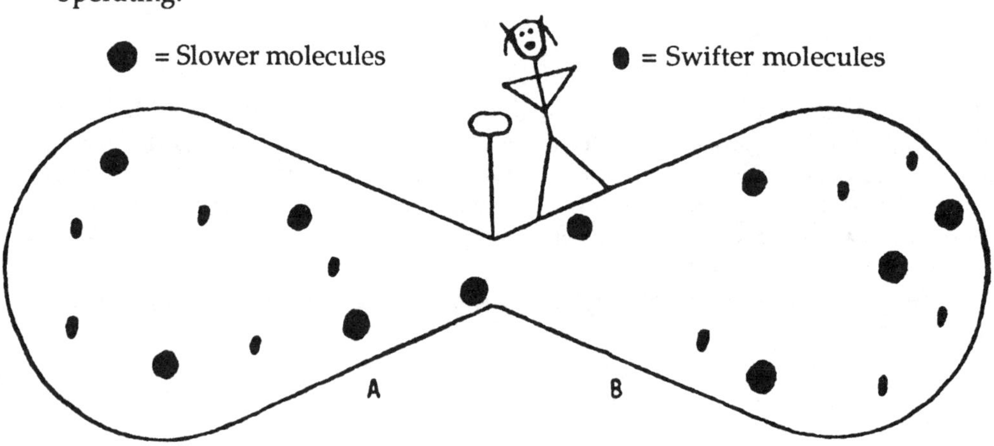

When the demon's task is completed chamber A is cooler than chamber B. There is more order since the total molecules inside the system are not randomly mixed. There is not an equilibrium throughout, and the chambers do not have equal temperatures.

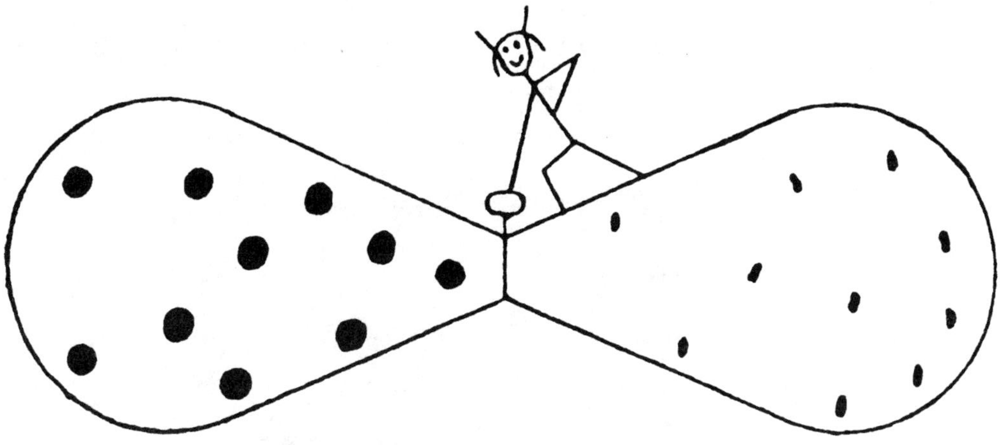

APPENDIX C: S = k log W

"In the above expression k is a numerical constant, and W is the number of physically distinct microscopic states of the system which correspond equally well to a given macroscopic state specified by known values of thermodynamic coordinates. The quantity W is often called the thermodynamic weight of a state which has been macroscopically characterized as having thermodynamic coordinates with given values. W is therefore the number of distinct microstates which agree with what is actually known about the state of the macroscopic system. The formula $S = k \log W$ for the statistical entropy of a system is essentially Boltzmann's definition of S. From this formula the first and second laws of thermodynamics are readily derived.

"The outstanding achievement represented by the Boltzmann formula $S = k \log W$ is that it defines the entropy S absolutely. No ambiguous additive constant is present in this formula once the thermodynamic weight W is uniquely defined. Classical thermodynamics offers no such possibility for a unique and absolute definition of the value of entropy since only the entropy difference $dS = dQ/T$ is defined at the level of macroscopic thermodynamics. Only entropy differences between distinct equilibrium states are defined in classical thermodynamics. The Boltzmann formula $S = k \log W$ allows an absolute determination of the value of the entropy in any equilibrium state."*

---

* Michael J. Harrison, "Entropy Concepts in Physics." Libor Kubat and Jiri Zeman, eds., *Entropy and Information in Science and Philosophy* (New York. Elsevier Scientific Publishing Company, 1975), pp. 46-47.

APPENDIX D: CHRONOLOGICAL LIST OF SCIENTISTS

(1642-1727)	Newton
(1686-1736)	Fahrenheit
(1728-1799)	Black
(1731-1810)	Cavendish
(1753-1814)	Rumford
(1796-1832)	Carnot
(1814-1878)	Mayer
(1818-1889)	Joule
(1821-1894)	Helmholtz
(1822-1888)	Clausius
(1824-1907)	Kelvin
(1831-1879)	Maxwell
(1839-1903)	Gibbs
(1844-1906)	Boltzman
(1887-1961)	Shrodinger
(1889-1969)	Brillouin
(1894-1964)	Wiener
(1898-1964)	Szilard
(1900-1985)	Lindsay
(1901-1972)	Bertalanffy
(1916-2001)	Shannon
(1917-2003)	Prigogine

BIBLIOGRAPHY ON ENTROPY

Allee, Warder Clyde. *The Social Life of Animals.* London: William Heinemann, 1955.

Ames, J.S., ed. *Harper's Scientific Memoirs VI. The Second Law of Thermodynamics.* New York: American Book Company, 1899.

Angrist, Stanley W., and Loren G. Hepler. *Order and Chaos: Laws of Energy and Entropy.* New York: Basic Books, 1967.

Arnheim, Rudolf. *Entropy and Art.* Berkeley: University of California Press, 1971.

_______. *Toward a Psychology of Art.* Berkeley: University of California Press, 1966.

Bent, Henry A. "Entropy & the Energy Crisis," *The Science Teacher,* 44 (May, 1977), 25-30.

_______. "Haste Makes Waste—Pollution and Entropy," *Chemistry,* 44 (October, 1971), 6-15.

Blum, Harold F. *Time's Arrow and Evolution.* Princeton: Princeton University Press, 1970.

Bragg, Sir William. *Concerning the Nature of Things.* New York: Dover, 1948.

Brillouin, L. *Science and Information Theory.* New York: Academic Press, 1956.

Brix, V. H. *You Are a Computer.* New York: Emerson Books, 1970.

Cannon, Walter B. *The Wisdom of the Body.* New York: Norton, 1963.

Chan, Wing-Tsit. *A Source Book in Chinese Philosophy.* Princeton: Princeton University Press, 1963.

Curtis, Helena. *Biology.* New York: Worth, 1975.

Davis, David E. *Integral Animal Behavior.* New York: Macmillan, 1967.

Dugdale, J.S. *Entropy and Low Temperature Physics.* London: Hutchinson & Co., 1966.

Eddington, Sir Arthur. *The Nature of the Physical World.* Ann Arbor: University of Michigan Press, 1963.

Einstein, Albert. *Out of My Later Years.* New York: The Philosophical Library, 1950.

Fast, John Diedrich. *Entropy.* New York: Gordon and Breach, 1968.

Fox, Fred W. "The Recolored Mentality—Ethical Lessons From Science," *The Science Teacher,* 43 (September, 1976), 22-24.

Fuller, R. Buckminster. *And it Came to Pass—Not to Stay.* New York: Macmillan, 1976.

_______. *No More Secondhand God*. New York: Doubleday, 1971.

_______. *Operating Manual for Spaceship Earth*. New York: Simon & Schuster, 1974.

_______.*Utopia or Oblivion: The Prospects for Humanity*. New York: Bantam Books, 1972.

Gabel, Medard. *Energy, Earth, and Everyone*. San Francisco: Straight Arrow Books, 1975.

Gamow, George. *Matter, Earth, and Sky*. New Jersey: Prentice-Hall, 1958.

Gatlin, Lila L. *Information Theory and the Living System*. New York: Columbia University Press, 1973.

Georgescu-Roegen, Nicholas. *The Entropy Law and the Economic Process*. London: Oxford University Press, 1971.

Gibbs, J. Willard. *Elementary Principles in Statistical Mechanics*. New York: Dover, 1960.

Hassenstein, Bernhard. *Information and Control in the Living Organism*. London: Chapman and Hall, 1971.

Kepes, Gyorgy, ed. *Structure in Art and Science*. New York: Braziller, 1965.

Klein, Ernest. *A Comprehensive Etymological Dictionary of the English Language*. New York: Elsevier, 1971.

Kohler, Wolfgang. *The Place of Value in a World of Facts*. New York: New American Library, 1966.

Kropotkin, Petr Alexekieevich. *Ethics Origin and Development*. New York: The Dial Press, 1924.

_______. *Mutual Aid: A Factor of Evolution*. New York: Alfred T. Knopf, 1914.

Kubat, Libor, and Jiri Zeman, eds. *Entropy and Information in Science and Philosophy*. New York: Elsevier, 1975.

Landsberg, P.T. *Entropy and the Unity of Knowledge*. Cardiff: University of Wales Press, 1961.

Lehninger, Albert L. *Bioenergetics*. London: W.A. Benjamin, 1973.

Lindsay, R.B. "Entropy Consumption and Values in Physical Science," *American Scientist*, 47 (September, 1959), 376-385.

_______. *The Nature of Physics*. Providence: Brown University Press, 1971.

_______. *The Role of Science in Civilization*. Westport: Greenwood Press, 1973.

Lovejoy, Arthur O. *The Great Chain of Being*. New York: Harper and Row, 1960.

Maxwell, James Clerk. *Theory of Heat*. London: Longman, Green and Co., 1883.

Meyer, Leonard B. *Music, the Arts, and Ideas.* Chicago: University of Chicago Press, 1967.

Moles, Abraham. *Information Theory and Esthetic Perception.* Urbana: University of Illinois Press, 1966.

Montagu, Ashley M.F. *The Direction of Human Development.* New York: Harper and Brothers, 195-5.

_______. *On Being Human.* New York: Hawthorn Books, 1967.

Morowtiz, Harold J. *Entropy for Biologists.* New York: Academic Press, 1970.

Needham, Joseph. *Order and Life.* Cambridge: The M.I.T. Press, 1968.

NSF Chautauqua-Type Short Courses for College Teachers, Washington, D.C.: Office of Science Education, American Association for the Advancement of Science, (Fall, 1977).

Oparin, Aleksander Ivanovich. *Life, its Nature, Origin, and Development.* New York: Academic Press, 1964.

Plank, Rudolf. *The Concept of Entropy.* Karlsruhe, Germany: Institute of Technology, 1962.

Polgar, Steven. "Evolution and the Thermodynamic Imperative," *Human Biology,* 33 (May, 1961), 99-109.

Potter, Van Rensselaer. *Bioethics.* New Jersey: Prentice-Hall, 1971.

Prigogine, I. *Introduction to Thermodynamics of Irreversible Processes.* New York: Interscience, 1955.

_______ and Robert Herman. *Kinetic Theory of Vehicular Traffic.* New York: Elsevier, 1971.

_______. *Non-equilibrium Statistical Mechanics.* New York: Interscience, 1962.

_______ ed. *Preceedings of the International Symposium on Transport Processes in Statistical Mechanics.* New York: Interscience, 1958.

_______. *Thermodynamics of Irreversible Processes.* New York: Interscience, 1961.

_______, Gregoire Nicolis, and Agnes Babloyantz. "Thermodynamics of Evolution," *Physics Today,* 25 (November, 1972), 23-28.

_______, Gregoire Nicolis, and Agnes Babloyanta. "Thermodynamics of Evolution," *Physics Today,* 25 (December, 1972), 38-44.

Procaccia, Itamar, and John Ross. "The 1977 Nobel Prize in Chemistry," *Science,* 198 (18 November, 1977), 716-717.

Quastler, Henry. *The Emergence of Biological Organization.* New Haven: Yale University Press, 1964.

Samuel, Edmund. *Order: In Life.* New Jersey: Prentice-Hall, Inc., 1971.

Saperstein, Alvin M. *Physics: Energy in the Environment*. Boston: Little, Brown and Company, 1975.

Sartwell, Joyce G., and Richard P. Abell. "A Teacher's Guide to the Energy Crisis," *Today's Education*, 64 (January-February, 1975), 90-97.

Shapley, Harlow. *Of Stars and Men*. Boston: Beacon Press, 1964.

Shrodinger, Erwin. *What Is Life?* Cambridge: Cambridge University Press, 1945.

Szilard, Gertrud Weiss, and Bernard T. Feld. eds. *The Collected Works of Leo Szilard: Scientific Papers*. Cambridge: The MIT Press, 1972.

Watson, Donald. "Energy Use and Building Design," *Science*, 198 (28 October, 1977), 391-392.

Watson, Fletcher G., F. James Rutherford, and Gerald Holton, directors. The *Project Physics Course, Reader 3 The Triumph of Mechanics*. New York: Holt, Rinehart and Winston, 1971.

Whyte, Lancelot Law, ed. *Apsects of Form*. London: Lund Humphries, 1968.

Wiener, Norbert. *Cybernetics*. New York: John Wiley and Sons, 1948.

_______. *The Human Use of Human Beings*. Boston: Houghton Mifflin, 1950.

Wilson, Alan Geoffrey. *Entropy in Urban and Regional Modelling*. London: Pion, 1970.

The Wolfcreek Statement: Toward a Sustainable Energy Society. Atlanta, Georgia: The Georgia Conservancy, (March, 1977).

Zeman, Jiri, ed. *Time in Science and Philosophy*. New York: Elsevier, 1971.

Zipf, George Kingsley. *Human Behavior and the Principle of Least Effort*. Cambridge: Addison-Wesley, 1949.

BIBLIOGRAPHY ON ETHICS

Aristotle. *Ethics*. New York: Oxford University Press, 1925.

Baier, Kurt. *The Moral Point of View: A Rational Basis of Ethics*. Ithaca: Cornell University Press, 1958.

Beck, Clive M., Brian S. Crittenden, and Edmund V. Sullivan, eds. *Moral Education: Interdisciplinary Approaches*. New York: Newman Press, 1971.

Blumenthal, Albert. *Moral Responsibility: Mankind's Greatest Need*. Santa Ana: Rayline Press, 1975.

Bowyer, Carlton H. *Philosophical Perspectives for Education.* Glenview: Scott, Foresman and Company, 1970.

Dewey, John. *Moral Principles in Education.* Boston: Houghton Mifflin, 1909.

Engel, David E., ed., *Religion in Public Education.* New York: Paulist Press, 1974.

Frankena, William K. *Ethics.* New Jersey: Prentice-Hall, 1963.

_______. "Moral Education, Philosophical View of," *The Encyclopedia of Education,* 1971, 6, 394-398.

Gallup, George. "The Public Looks at the Public Schools," *Today's Education,* 64 (September-October, 1975), 16-20.

_______ and Vernon Smith. *What the People Think About Their Schools: Gallup's Findings.* Bloomington: Phi Delta Kappa Educational Foundation, 1977.

Harris, Thomas A. *I'm OK—You're OK.* New York: Avon Books, 1973.

Journal of Thought, 12 (January, 1977), The entire issue.

Kohlberg, Lawrence. "Moral Education, Psychological View of," *The Encyclopedia of Education,* 1971, 6, 398-406.

Lerner, Max. *Values in Education.* Bloomington: Phi Delta Kappa Educational Foundation, 1976.

McGuire, Willard. "What Can We Do About Violence?" *Today's Education,* 64 (November-December, 1975), 22-23.

Peters, Richard S. *Ethics and Education.* Chicago: Scott, Foresman, 1966.

Phi Delta Kappan, 56 (June, 1975), The entire issue.

_______, 59 (January, 1978), The entire issue.

Piaget, J. *The Child's Conception of the World.* New York: Harcourt, Brace & Company, 1929

_______. *The Construction of Reality in the Child.* New York: Basic Books, 1954.

Raths, Louis E, Merrill Harmin, and Sidney B. Simon. *Values and Teaching.* Columbus: Charles E. Merrill, 1966.

Simon, Sidney B., Leland W. Howe, and Howard Kirschenbaum. *Values Clarification.* New York: Hart, 1972.

Wilson, John, Norman Williams, and Barry Sugarman. *Introduction to Moral Education.* Baltimore: Penguin, 1967.

The life impulse manifested in growth and development at the beginning of life and in adulthood constitutes a major countertendency to this second law of thermodynamics. (*entropy*)

—Hugh Perkins

Man's physical body and his social aspirations are anti-entropic.

—Fred Fox

All men should fight always as vigorously as possible to increase the degree of order in their environment, i.e., consume as much entropy as possible, in order to combat the natural tendency for entropy to increase and for order in the universe to be transformed into disorder, in accordance with the second law of thermodynamics.

—Robert Lindsay

Until now life has succeeded in maintaining itself: its persistence has retarded entropy.

—Steven Polgar

... the fact that during growth living organisms actually show a decrease of entropy production during evolution up to the stationary state. Also, the fact that their organization generally increases during this evolution corresponds to the decrease of entropy as studied.

—Ilya Prigogine

...the living organism "...feeds on negative entropy."

—Erwin Schrodinger

What we don't need is a national energy policy. What we do need is a *personal entropy ethic.*

—Henry Bent

Since stored information varies inversely with entropy, lowered entropy means a higher capacity to store information.

—Lila Gatlin

The small local decrease in entropy represented in the building of the organism is coupled with a much larger increase in the entropy of the universe.

—Harold Blum

Each localized, man-made or machine-made entropy decrease is accompanied by a greater increase in entropy of the surroundings, thereby maintaining the required increase in total entropy.

—Angrist and Hepler

In order to prevent ourselves from dissipating to an equilibrium state of death, we require a constant flow-through of free energy (negative entropy) from our larger environment.

—Jeremy Rifkin

Once it is understood that human beings are "one" with nature, then an ethical base is established by which the appropriateness of all human activity can be judged ... In a low-entropy society the notion of "conquering" nature is replaced by the idea of harmony with other creatures and the environment as a whole.

—Rifkin

EPILOGUE

It was twenty years ago this semester when Vy Carswell first asked me to visit her classroom. We had talked about the best way to teach middle school students how to classify energy. I was pleased to have such a vivacious, witty, intelligent, pretty woman invite me to observe her interactions with students. I accepted.

Vy's science lesson that day was to have her students think about why the vanes in a radiometer moved when exposed to light (photons of electromagnetic radiation). A solar energy arrow was used from the sun to the vanes in the vacuum. Then a curved arrow, labeled mechanical energy, indicated the circular motions if the vanes. For us, using arrows became the best way for students to understand and remember energy classifications and energy flow conceptualization.

Vy is a Master Teacher who finished her Masters Degree. She had been teaching science for more than 15 years, in the only 6th grade middle school in San Marcos, Texas. The town had about 15,000 people when Vy started. When I met her, she had helped more than 1,500 students through their first full year of science and the town had grown to about 21,000; 67% Hispanic. Vy's students and their parents all loved her, and they still do. Me, too!!!

Three years later, we married. To the old timers here, and their grown children, I am still known as "Mis Carswell's husband." She leaves a powerful, happy, positive and respectful memory with her students, and she never forgets their names. I am proud she invited me into her life.

Professor Victor Michalk gave up a summer of travel to help us with grade 1-12 teachers. For many of them, it was their first time in any graduate or even physical science undergraduate course. It was easy for Victor to show how fusion and fission happened, and many other energy transformations. Victor and Vy complemented each other while helping the students with patience and compassion. Relaxed, calm, and with wit, they guided the students toward the new curriculum and two well done booklets they would build and turn in to the U.S. Department of Education.

The covers of the 1988 booklets are introduced in the appendix of this book. The first booklet was conceptualized by Helen Lawrence's students; and the second was conceptualized by Bobby Jones. Bobby was the only science teacher from one of the smallest school districts in Texas–Utopia, Texas, which graduated 9 to 12 students per year.

Grant clerks Beverly Rodriguez and Jeff Spencer were students in financial accounting. As any grant receiver knows, these students are very valuable.

Carol Dolezal received a grant to ink original illustrations for film and photocopy mass reproduction. Carol, a well known art teacher in our county school district and admired for her painting, went through the physics course and then worked her wonders. An example from each of the five sources of energy follows this epilogue.

Carol Dolezal

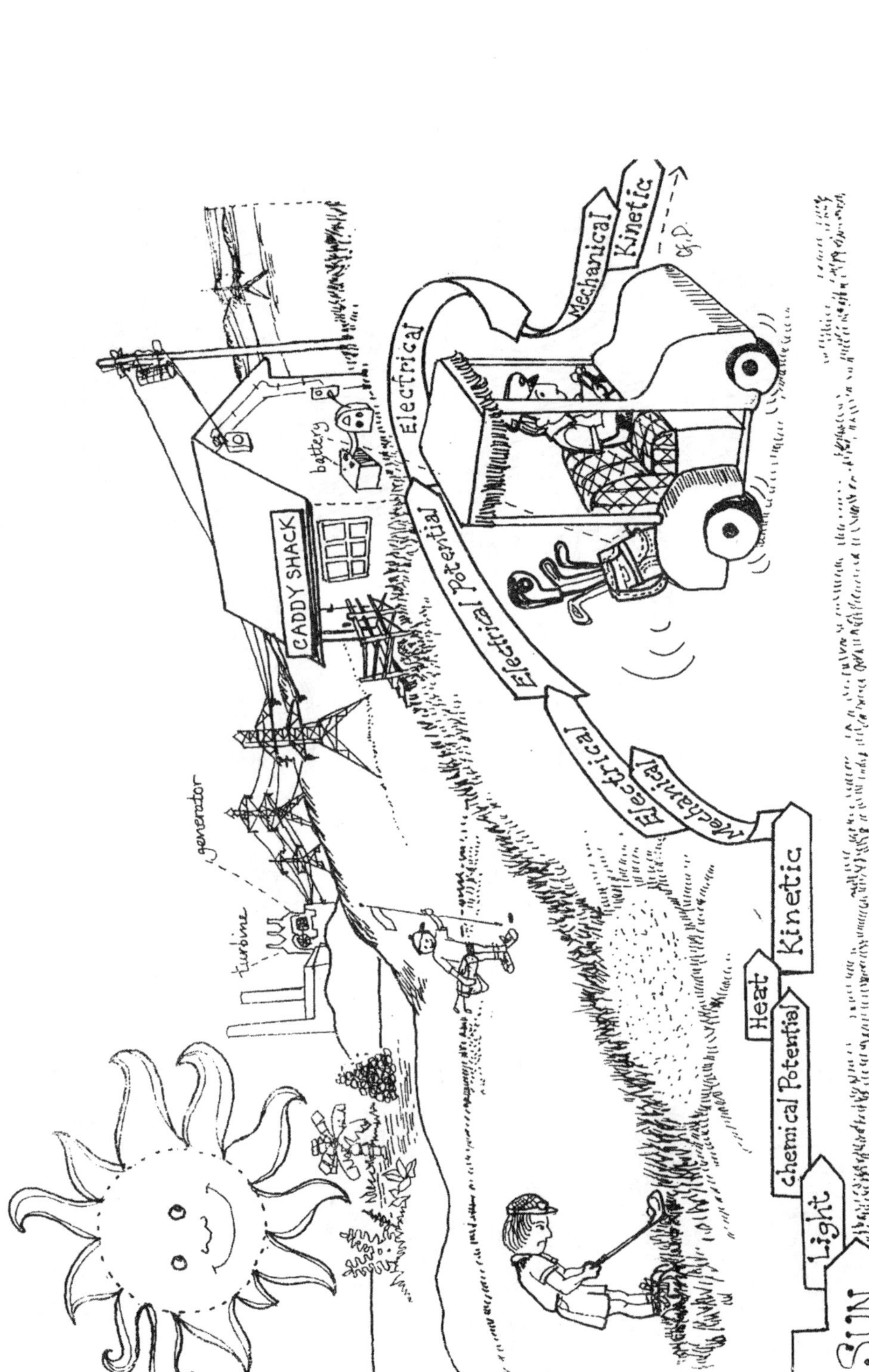

Carol Dolezal

Carol Dolezal

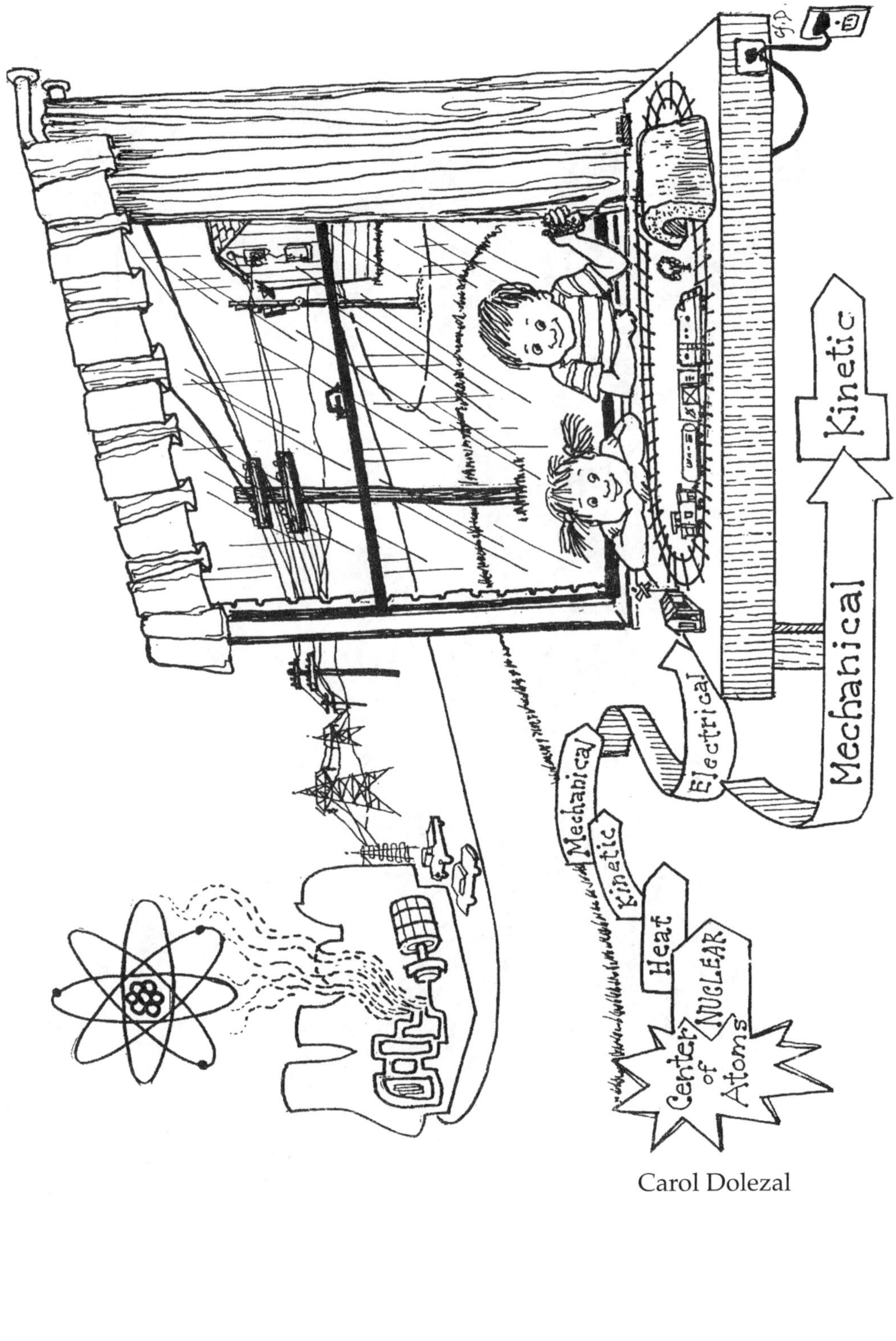

Carol Dolezal

Carol Dolezal

C - D E E P

In the early 1980s the State of Texas was asking all fourth grade teachers to have their students learn the classifications of energy, matter and the forces in nature. No other instructions were given.

I designed a survey form, asking all of the school districts around our campus which of the three gave them the most trouble. The fourth grade teachers asked for help in classifying energy. After meetings with our Department of Physics and the Seguin Independent School District, I began the grant writing. The acronym C-DEEP was agreed to by all. It became synonymous with "The Seguin Project".

Ten years earlier, I thought of using C-DEEP on my doctoral dissertation; "Curriculum Development in the Entropy Ethic Project". My committee wanted something more scholarly (I have come to know that is not unusual). Thus my first drawing of THE HUMAN SYSTEM appears in my copyrighted dissertation, "An Analysis Of Entropy Reduction And Its Implications For Ethical Instruction In Public Education".

From 1985 through 1995, C-DEEP went national and international as "Curriculum Development in the Energy Education Project". Science teachers K through 12th grades, using poster board, cut color-coded and shape-coded arrows for the sources of energy, the forms of energy and the ways to store energy. This was the same process they would teach their students. They could write anything they wanted on their arrows. Examples of their work appear in the Eulogy Edition Special Collection at the end of this book.

The teachers and their students soon learn to conceptualize energy flow, another unseeable theory – they must go up the ladder of inductive reasoning and down the ladder of deductive reasoning to understand and to know. Then they become prepared for systems analyses, the age of information, and THE ENTROPY ETHIC.

Funding for C-DEEP came from:

- A grant from Phi Delta Kappa (1982-1983)
- Four grants from the LBJ Institute for the Improvement of Teaching and Learning (1988, 1990)
- Two federal grants from the Education for Economic Security Act (U. S. Department of Education) through the Texas Higher Education Coordination Board (1988)
- A grant from the Guadalupe-Blanco River Authority (1990)

C - D E E P

Curriculum **D**evelopment in **E**nergy **E**ducation **P**roject

OUTLINE FOR

Energy Education • Graduate Course • Computer Program • Book

INTRODUCTION

PART I. Basic Binding Forces in Nature

Chapters
1. Gravity
2. Electromagnetic
3. Strong Nuclear
4. Weak

PART II. The Sources of Energy

Chapters
5. Sun
6. Center of Atoms
7. Center of Earth
8. Rotation of Earth
9. Gravity
10. Other Stars

PART III. The Forms of Energy

Chapters
11. Heat
12. Electricity
13. Light
14. Mechanical
15. Muscular
16. Motion
17. Sound

PART IV. The Ways to Store Energy

Chapters
18. Chemical
19. Electrical
20. Gravitational
21. Mechanical

PART V. Energy Flow Through Systems

Chapters
22. Isolated
23. Closed
24. Open

PART VI. Energy and Information Laws

Chapters
25. Energy #1
26. Energy #2
27. Information #1
28. Information #2
29. Information #3
30. Information #4

PART VII. Energy Flow Conceptualization

CONCLUSION

ENERGY DISSIPATING DEVICES

BIBLIOGRAPHY

Southwest Educational Development Laboratory

and its

Rural, Small Schools Initiative

recognizes the

Curriculum Development in Energy Education Project (C-DEEP)

at

Seguin Independent School District

as an Exemplary Program in the use of
Systematic Staff Development for School Improvement

Summer, 1990

Preston C. Kronkosky, Executive Director, SEDL

Marianne Vaughan, Assistant Executive Director
Services for School Improvement, SEDL

A CURRICULUM ACTIVITIES GUIDE

FOR IMPROVING INSTRUCTION IN 5TH AND 6TH GRADE PHYSICS

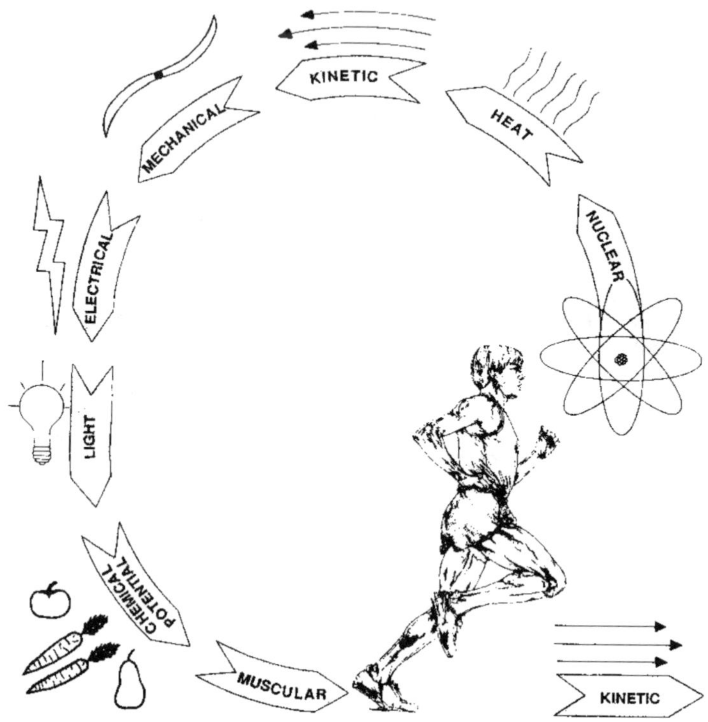

Nuclear - Powered Jogger

Concept by H. Lawrence

Under the guidance of Professors Dick E. Hammond, Department of Curriculum and Instruction, and Dr. Victor E. Michalk, Department of Physics, with assistance of Vy Carswell and Beverly Rodriguez.

Southwest Texas State University - San Marcos

A CURRICULUM ACTIVITIES GUIDE

FOR IMPROVING THE TEACHING AND LEARNING OF PHYSICS

concept by Bobby Jones

Under the guidance of Professors Dick E. Hammond, Department of Curriculum and Instruction, and Dr. Victor E. Michalk, Department of Physics, with graduate assistant Vy Carswell-Hammond and Grant Clerk Jeff Spencer.

Southwest Texas State University

SCIENTIFIC KNOWLEDGE REVOLUTIONS

	1400	Renaissance
#1	**1500**	First Scientific Revolution (Copernicus)
	1600	Society of Jesus (written exams)
	1700	Newton (Universal Law of Gravity)
	1800	Industrial Revolution
	1850s	Second Scientific Revolution
#2	**1864**	Texas A&M
	1900	(Southwest) Texas State University College Entrance Exams
	1905	Binet Mental Test (Special Theory of Relativity)
	1917	Army Examination
	1947	Educational Testing Service
#3	**1977**	Systems Analysis, Third Scientific Revolution, Information Theory, DNA...Ilya Prigogine's Nobel Prize
	1980	Rifkin, Jeremy. *Entropy: A New World View*
	1984	von Bertalanffy, Ludwig. *General System Theory* (9th printing) Prigogine, Ilya and Stengers, Isabelle. *Order Out Of Chaos*
	1994	Leff, Harvey S. and Rex, Andrew F., editors. *Maxwell's Demon: Entropy Information Computing* Csikszentmihalyi, Mihaly. *The Evolving Self: A Psychology For The Third Millenium*
	2002	Danchin, Antoine. *The Delphic Boat: What Genomes Tell Us* Rifkin, Jeremy. *The Hydrogen Economy*
	2003	Prigogine dies
	2005	Hammond, Dick E. *The Human System From Entropy To Ethics*

Information Is Inversely Related To Entropy.

A letter is better when straightened around.
It's hard to read when it's upside down.
What if a...B ...were laid in its back?
Two little hills and one middle crack.

Back is harder to read this way.
It's harder to see and harder to say.
And what about... C ...Now there's a letter,
It could roll on its back... ...but isn't this better... C

We don't write our letters this way...

...or that.
It's too hard to read a word like...CAT.

What if the letters were flying in air?
Some were here... and some were there.
That gives us no picture of CAT or kitten,
When the letters for CAT are wrongly written.

Or what if the letters took long water rides
Bounding and bumping on each other's sides.

Wouldn't the CAT be harder to know
Than with all the letters placed in a row? CAT

DEPT. OF
ENTROPY
1967

Dr Hammond
Human Growth and Development II
Fall... 1995
Entropy and the Human System